Lecture Notes in Computer Science 16256

Founding Editors

Gerhard Goos
Juris Hartmanis

Editorial Board Members

Elisa Bertino, *Purdue University, West Lafayette, IN, USA*
Wen Gao, *Peking University, Beijing, China*
Bernhard Steffen, *TU Dortmund University, Dortmund, Germany*
Moti Yung, *Columbia University, New York, NY, USA*

The series Lecture Notes in Computer Science (LNCS), including its subseries Lecture Notes in Artificial Intelligence (LNAI) and Lecture Notes in Bioinformatics (LNBI), has established itself as a medium for the publication of new developments in computer science and information technology research, teaching, and education.

LNCS enjoys close cooperation with the computer science R & D community, the series counts many renowned academics among its volume editors and paper authors, and collaborates with prestigious societies. Its mission is to serve this international community by providing an invaluable service, mainly focused on the publication of conference and workshop proceedings and postproceedings. LNCS commenced publication in 1973.

Sergio Cavero · Eduardo G. Pardo ·
Angelo Sifaleras
Editors

Variable Neighborhood Search

11th International Conference, ICVNS 2025
Montreal, QC, Canada, May 12–14, 2025
Revised Selected Papers

 Springer

Editors
Sergio Cavero
Universidad Rey Juan Carlos
Madrid, Spain

Eduardo G. Pardo
Universidad Rey Juan Carlos
Madrid, Spain

Angelo Sifaleras
University of Macedonia
Thessaloniki, Greece

ISSN 0302-9743   ISSN 1611-3349 (electronic)
Lecture Notes in Computer Science
ISBN 978-3-032-19581-4   ISBN 978-3-032-19582-1 (eBook)
https://doi.org/10.1007/978-3-032-19582-1

This Springer imprint is published by the registered company Springer Nature Switzerland AG
The registered company address is: Gewerbestrasse 11, 6330 Cham, Switzerland

If disposing of this product, please recycle the paper.

Preface

This volume contains peer-reviewed papers from the 11th International Conference on Variable Neighborhood Search (ICVNS 2025), held in Montreal, Canada, from May 12 to 14, 2025. The conference was jointly organized within the *Journées de l'Optimisation* (Optimization Days) by the Interuniversity Research Centre on Enterprise Networks, Logistics and Transportation (CIRRELT) at HEC Montréal, Canada.

The conference follows previous successful meetings that were held in Abu Dhabi, UAE (2022 and 2021); Rabat, Morocco (2019); Sithonia, Halkidiki, Greece (2018); Ouro Preto, Brazil (2017); Málaga, Spain (2016); Djerba, Tunisia (2014); Herceg Novi, Montenegro (2012); and Puerto de La Cruz, Tenerife, Spain (2005).

This edition holds profound emotional significance for our community, serving as a tribute to Pierre Hansen (1940–2025), the pioneer of Variable Neighborhood Search, who passed away on January 19, 2025. Along with Nenad Mladenović (1951–2022), Pierre Hansen founded this metaheuristic and served as the Honorary Chair of this conference series for many years. We deeply mourn his passing, yet we celebrate his intellectual generosity and passion for research, and the indelible mark he left on the field of optimization. His legacy, alongside that of Nenad Mladenović, continues to inspire the research presented in this volume.

The main goal of ICVNS 2025 was to provide a stimulating environment where researchers, practitioners, and experts in optimization, heuristics, and metaheuristics could discuss the latest advancements in VNS and its applications in machine learning, logistics, scheduling, and graph problems, among others.

The plenary speaker of ICVNS 2025, Jack Brimberg, from the Royal Military College, Kingston, Canada, presented to the audience his talk *"Variable Neighborhood Search: Early days to today"*, where he offered a historical journey through VNS, tracing the origins of the methodology to 1995, when it was presented for the first time at the *Journées de l'Optimisation*, and discussing its evolution up to current trends.

Around 40 participants took part in the ICVNS 2025 conference, and among the presented works, 14 papers were selected for publication in this LNCS volume from 20 submissions. The selection process was rigorous; each paper underwent a single-blind peer-review process involving at least two reviews per paper by members of the Program Committee. These papers describe recent advances in the methodology of VNS as well as novel applications in combinatorial and global optimization.

The editors thank all the participants for their contributions and their continuous effort to disseminate VNS and are grateful to the reviewers for preparing excellent and detailed reports. The editors wish to acknowledge the Springer LNCS editorial staff for their support during the entire publication process.

Finally, we express our gratitude to the organizers and sponsors of the ICVNS 2025 meeting, especially:

- GERAD (Group for Research in Decision Analysis),

- CIRRELT (Interuniversity Research Centre on Enterprise Networks, Logistics and Transportation),
- HEC Montréal,
- The organizers of the *Journées de l'Optimisation* (JOPT).

Their support was essential in making ICVNS 2025 a memorable scientific event honoring the memory of our founders.

May 2025

Sergio Cavero
Eduardo G. Pardo
Angelo Sifaleras

Organization

Conference Chairs

Daniel Aloise	Polytechnique Montréal, Canada
Rachid Benmansour	INSEA, Morocco
Tatjana Davidović	Mathematical Institute, Serbian Academy of Sciences, Serbia
José Andrés Moreno Pérez	Universidad de La Laguna, Spain
Eduardo G. Pardo	Universidad Rey Juan Carlos, Spain
Angelo Sifaleras	University of Macedonia, Greece
Dragan Urošević	Mathematical Institute, Serbian Academy of Sciences, Serbia

Program Committee

Daniel Aloise	Polytechnique Montréal, Canada
Cláudio Alves	University of Minho, Portugal
Ada Álvarez	Universidad Autónoma de Nuevo León, Mexico
John Beasley	Brunel University London, UK
Rachid Benmansour	INSEA, Morocco
Jack Brimberg	Royal Military College, Canada
Lucidio Cabral	Universidade Federal da Paraíba, Brazil
Gilles Caporossi	HEC Montréal, Canada
Emilio Carrizosa	University of Seville, Spain
Alejandra Casado Ceballos	Universidad Rey Juan Carlos, Spain
Sergio Cavero	Universidad Rey Juan Carlos, Spain
José Manuel Colmenar	Universidad Rey Juan Carlos, Spain
Sergio Consoli	European Commission, Joint Research Centre, Belgium
Teodor G. Crainic	Université du Québec à Montréal, Canada
Tatjana Davidović	Mathematical Institute SANU, Serbia
Karl Dörner	University of Vienna, Austria
Adriana Gabor	Khalifa University, United Arab Emirates
Michel Gendreau	École Polytechnique de Montréal, Canada
Sergio Gil Borrás	Universidad Politécnica de Madrid, Spain
Saïd Hanafi	Université Polytechnique des Hauts-de-France, France

Richard Hartl   University of Vienna, Austria
Chandra Irawan   Nottingham University Business School, China
Bassem Jarboui   Université Polytechnique des Hauts-de-France,
                                          France
Raja Jayaraman   New Mexico State University, USA
Yuri Kochetov   Sobolev Institute of Mathematics, Russia
Vera Kovačević-Vujčić   University of Belgrade, Serbia
Mário Leite   University of Minho, Portugal
Isaac Lozano-Osorio   Universidad Rey Juan Carlos, Spain
Yannis Marinakis   Technical University of Crete, Greece
Lucas Martín-García   Universidad Rey Juan Carlos, Spain
Raúl Martín-Santamaría   Universidad Rey Juan Carlos, Spain
Anna Martínez-Gavara   University of Valencia, Spain
Fouad Medouar   INSEA, Morocco
Mahuton Hugues Midingoyi   Polytechnique Montréal, Canada
José A. Moreno-Pérez   University of La Laguna, Spain
Rustam Mussabayev   National Institute of Information Security
                                          Development, Kazakhstan
Vitor Nazário Coelho   Seiva Computação, Brazil
Luiz Satoru Ochi   Fluminense Federal University, Brazil
Panos Pardalos   University of Florida, USA
Eduardo G. Pardo   Universidad Rey Juan Carlos, Spain
Jun Pei   University of Hefei, China
Iván Penedo   Universidad Rey Juan Carlos, Spain
Sergio Pérez-Peló   Universidad Rey Juan Carlos, Spain
Filipe Pessoa Sousa   Universidade do Estado do Rio de Janeiro, Brazil
Leonidas Pitsoulis   Aristotle University of Thessaloniki, Greece
Günther Raidl   TU Wien, Austria
Varvara Rasskazova   Moscow Aviation Institute, Russia
Miguel Reula   University of Valencia, Spain
Celso Ribeiro   Universidade Federal Fluminense, Brazil
Marcos Robles   Universidad Rey Juan Carlos, Spain
Sergio Salazar Cárdenas   Universidad Rey Juan Carlos, Spain
Said Salhi   University of Kent, UK
Jesús Sánchez-Oro   Universidad Rey Juan Carlos, Spain
Joan Serrano Roig   University of Valencia, Spain
Marc Sevaux   University of Southern Brittany, France
Patrick Siarry   Université Paris-Est Créteil Val de Marne, France
Angelo Sifaleras   University of Macedonia, Greece
Andrei Sleptchenko   Khalifa University, United Arab Emirates
Kenneth Sörensen   University of Antwerp, Belgium
Marcone Jamilson Freitas Souza   Federal University of Ouro Preto, Brazil

Zorica Stanimirović	University of Belgrade, Serbia
Christos D. Tarantilis	Athens University of Economics and Business, Greece
Raca Todosijević	Université Polytechnique des Hauts-de-France, France
Michael-Alexandros Triantoglou	University of Macedonia, Greece
Hasan Turan	University of New South Wales, Australia
Dragan Urošević	Mathematical Institute SANU, Serbia
Silvia Ventura Cabrejas	Universidad Rey Juan Carlos, Spain
Javier Yuste	Universidad Rey Juan Carlos, Spain
Qiuhong Zhao	Beihang University, China

Organizing Committee

Daniel Aloise	Polytechnique Montréal, Canada
Maryam Darvish	Polytechnique Montréal, Canada
Fausto Errico	Polytechnique Montréal, Canada
Jorge E. Mendoza	Polytechnique Montréal, Canada

Contents

Solving the Minimum Positive Influence Dominating Set Problem in Social Networks Using Metaheuristics

Iván Penedo[1] , Isaac Lozano-Osorio[1(✉)] , Jesús Sánchez-Oro[1] , and Óscar Cordón[2]

[1] Department of Computer Science and Statistics, Universidad Rey Juan Carlos, Móstoles, Spain
`{ivan.penedo,isaac.lozano,jesus.sanchezoro}@urjc.es`
[2] Department of Computer Science and Artificial Intelligence, University of Granada, Granada, Spain
`ocordon@decsai.ugr.es`

Abstract. The rise of the internet and social networks has posed new challenges in studying people's behavior on these platforms. People tend to trust or align with a small group of users, leading to the development of viral marketing techniques to effectively propagate information about products or services. This has led to the definition of problems related to social influence maximization/minimization and dominance sets. The Minimum Positive Influence Dominating Sets (MPIDS) problem involves finding a minimum cardinality dominance set to influence an entire social network. For a vertex to be influenced, at least half of its neighbors must be in the dominance set. Considering that MPIDS is an $\mathcal{NP}$-hard problem, where exact approximations are impractical due to the size of social networks, this work proposes using Basic Variable Neighborhood Search (BVNS). Given an initial solution generated by a constructive method, this metaheuristic consists of two phases: a shaking and an improvement phase. In the shaking phase, the solution is modified by removing and reconstructing it using a randomized greedy approach. The improvement phase involves an innovative local search strategy based on generating holes, which removes the δ-neighborhood of a vertex to facilitate a greedy solution reconstruction.

Keywords: Social network influence · Metaheuristics · Basic Variable Neighborhood Search · Minimum Positive Influence Dominating Sets

1 Introduction

Social Influence Maximization (SIM) problems have been extensively studied in the context of viral marketing, where consumers sequentially influence their social relationships to purchase a product [4]. Some related work shows that interactions between users in a social network can be used to prevent the spread

S. Cavero et al. (Eds.): ICVNS 2025, LNCS 16256, pp. 1–14, 2026.
https://doi.org/10.1007/978-3-032-19582-1_1

of diseases [8], conduct viral marketing campaigns [11], show recommendations in e-learning software [17], study social relationships [7], and prevent tobacco or other substance abuse [25]. For these problems, we seek to identify those users who accelerate or reduce the diffusion of influence in the network [12,13,21].

The Minimum Positive Influence Dominating Set (MPIDS) seeks the smallest set of users to influence an entire social network. This problem was formally defined in [26] where it was shown to be an $\mathcal{NP}$-hard problem for general graphs. Also, in the scope of these problems, a social network is formally modeled as an undirected graph $G = (V, E)$. Each edge $(u, v) \in E$ has two ends u and v, indicating that these users are connected, hence one user has a relationship with another and can influence him/her in the represented social network. Furthermore, we define $N(v)$ as the set of vertices adjacent to v, i.e., $\{u \in V : (u, v) \in E\}$.

Given an undirected graph $G = (V, E)$, a Positive Influence Dominating Set (PIDS) tries to find a set of vertices $D \subseteq V$ such that at least half of the neighbors of any vertex lie in D, i.e., $|D \cap N(v)| \geq |N(v)|/2 \ \forall v \in V$. Specifically, the goal of the MPIDS problem is to obtain a dominant set of positive influence with minimum cardinality. Lin et al. [10] defined the following linear integer programming model 1.

$$\min \sum_{i=1}^{|V|} x_i$$

$$\text{s.t.} \sum_{v_j \in N(v_i)} x_j \geq \left\lceil \frac{|N(v_i)|}{2} \right\rceil \quad \forall i \in V \tag{1}$$

$$x_i \in \{0, 1\} \qquad \forall i \in V$$

Binary variables $x_i \in \{0, 1\}$ are assigned to each vertex $i \in V$, where $x_i = 1$ indicates that vertex i is selected, while $x_i = 0$ implies the opposite. On the other hand, the main constraint forces any feasible solution to have the necessary vertices selected so that all vertices $v_i \in V$ have at least half of their neighbors selected in D.

The main drawback is the size of today's social networks, which makes mathematical models unable in some cases to even provide a feasible solution. For this reason, there are different greedy or heuristic approaches in the literature.

First, Wang et al. [26] proposed a greedy algorithm that, at each iteration, selected the vertex that could influence a larger number of vertices, with an algorithmic complexity of $O(|V|^3)$. Later, Raei et al. [20] defined the parameter *cover-degree* to prioritize certain vertices for the greedy criteria reducing the time complexity to $O(|V|^2)$. Subsequently, the Fast Greedy Algorithm (FGA) presented by Pan et al. [19] followed the same greedy strategy but propagating through the neighborhood of the last dominated vertex and prioritizing dominance over vertices that do not satisfy the constraint, reducing the complexity to $O(|V| \log |V| + |E|)$. The Improved Greedy Algorithm (IGA-PIDS) [3] is based on FGA and uses the parameter *need-degree* in addition to *cover-degree* to prioritize vertices. It also performs a pruning of the network at the beginning to obtain those vertices that must always be dominated, and a final sieve to not dominate redundant vertices in the dominance set at the end of the algorithm execution. The latter proposal obtains higher quality solutions than FGA, while maintaining a time complexity of $O(|V| + |E|)$.

On the other hand, different metaheuristic algorithms for the problem can be found in the literature. One of the first proposals was the ILPMA memetic algorithm by Lin et al. [10], which presented the mathematical formulation of the problem, in addition to proposing a Tabu Search to perform local optimizations on the dominance set. The first variant of a Construct, Merge, Solve and Adapt (CMSA) algorithm, proposed by Akbay and Blum [1], generated several solutions, then mixed them and performed several iterations using auxiliary adaptive structures to vary the generation of dominance sets. These structures were replaced by a variable n_a denoting the number of solutions generated in each iteration, which was incremented upon finding a solution with the same value of the objective function and reset upon finding a better solution in a later self-adaptive variant of that algorithm [2]. Then, the Iterated Carousel Greedy (ICG) [23] algorithm replaced the Tabu Search of [10] with an iterative process of destruction and reconstruction of the dominance set. Finally, the FastPIDS algorithm of Sun et al. [24] proposes several reduction rules for dominance set generation, as well as reusing the *need-degree* function from the literature. Also, the algorithm combines a greedy construction using a hybrid criteria with a local search based on vertex swapping, combined with a *two-level satisfaction judgment* mechanism. This mechanism uses two propositions to learn which partial dominance sets are promising and which are not to add new vertices to this set using the hybrid heuristic. The FastPIDS algorithm is currently the best performing algorithm according to the literature, so we will use it to compare against our proposal.

This paper presents a heuristic approach to provide quality solutions to the MPIDS problem with reduced computational time. A Basic Variable Neighborhood Search (BVNS) algorithm [18] has been designed, proposing a novel improvement based on creating holes in the solution and then applying an intelligent reconstruction of the dominance set.

The article is organized as follows. The design of the proposal is carried out in Sect. 2, which also describes the constructive stage and the improvement phase using the local search process. Section 3 presents the results obtained in the different experiments performed. These will be compared with the results of the best existing algorithm for the MPIDS problem to make a fair comparison. Finally, Sect. 4 mentions the conclusions reached with the development of this work and future work.

2   Basic Variable Neighborhood Search

The Variable Neighborhood Search (VNS) metaheuristic was proposed by Mladenović and Hansen in 1997 [18] as a systematic approach to explore multiple neighborhoods in optimization problems. Its main contribution lies in two key ideas: considering diverse neighborhood structures during the search and systematically switching between them to escape local optima. The algorithm alternates

4   I. Penedo et al.

phases of exploration (shaking) and exploitation (local search), progressively expanding the search radius. Thanks to this strategy, VNS effectively balances intensification and diversification, allowing it to explore promising regions of the solution space efficiently.

The pseudocode corresponding to the BVNS metaheuristic is presented in the Algorithm 1 where increasing neighborhoods are reached through the Shake operator, ensuring systematic diversification [6]. It requires as input the network G, together with the parameters α, δ (refer to the descriptions below), the maximum execution time t_{max}, and the maximum number of neighborhoods k_{max}. The procedure starts with a trivial initial solution that contains all vertices V (line 1). Then, a main loop is executed as long as the time limit is not reached (line 2).

In each iteration, an initial solution is generated through the `Construct` procedure, which uses α as a randomness parameter (line 3). Subsequently, this solution is improved by means of a local search with depth vertex removal parameter δ (line 4) and the neighborhood index k is initialized (line 5). From this solution, the characteristic BVNS process is repeated until variable k reaches the value of parameter k_{max} (line 6): increasing neighborhoods are reached through the `Shake` operator, which perturbs the current solution according to the neighborhood index k (line 7). Then, local search is applied again to intensify the generated solution (line 8) and it is decided whether to advance to a more distant neighborhood or to restart from the first one (line 9).

This process allows us to escape from local optima in a systematic way by exploring different neighborhood structures. When the algorithm reaches $k > k_{\max}$, if the solution D obtained in the BVNS process is better than the best known solution D_b (line 11), D_b is updated with the new dominance set (line 12). Finally, after running out of execution time, the best solution found D_b is returned (line 15).

Algorithm 1. $BVNS\ (G, \alpha, \delta, t_{\max}, k_{\max})$

1: $D_b \leftarrow V$
2: **while** $time \leq t_{\max}$ **do**
3:     $D \leftarrow Construct(G, \alpha)$
4:     $D \leftarrow LocalSearch(D, \delta)$
5:     $k \leftarrow 1$
6:     **while** $k \leq k_{\max}$ **do**
7:         $D' \leftarrow Shake(D, k, \alpha)$
8:         $D' \leftarrow LocalSearch(D', \delta)$
9:         $k \leftarrow NeighborhoodChange(D, D', k)$
10:     **end while**
11:     **if** $|D| < |D_b|$ **then**
12:         $D_b \leftarrow D$
13:     **end if**
14: **end while**
15: **return** D_b

Since the constructive process may cause the proposed solution to contain vertices that are not necessary to satisfy the constraints of the problem, a refinement process has been added. This process runs through the vertices added to the dominance set to check which ones can be removed while maintaining a feasible solution. For a vertex d to be considered redundant and therefore removed from the solution, the following constraint must be satisfied:

$$|N(v) \cap D| > \left\lceil \frac{N(v)}{2} \right\rceil \forall v \in N(d)$$

Therefore, the method will go through all the vertices belonging to the solution and eliminate those that meet the above restriction.

2.1 Constructive Procedure

The constructive phase seeks to obtain an initial solution starting from an empty solution and using a greedy criteria to select the candidate vertices to add to the dominance set.

The criteria followed in this work is based on the degree $|N(v)|$ of each of the vertices $v \in V$. In the constructive phase, it is necessary to evaluate the contribution of each of its vertices with a greedy criteria. Influence problems in social networks use datasets with millions of users, for that reason, a simple greedy construction is used for generating an initial solution before the BVNS procedure. Therefore, the constructive procedure iteratively adds the vertex with the largest degree to the incumbent solution until it becomes feasible.

Before this, it is important to note that some of the reductions proposed by Sun et al. [24] have been applied, which allow us to determine whether certain vertices should belong or not to the solution for different characteristics of the vertices shown below. The first reduction shows that, if a leaf vertex has only one parent vertex, in order to obtain a feasible solution it is necessary that the parent is always in the set D. Meanwhile, in case of having a vertex structure forming a triangle between the relations, it will be necessary that the solution D contains two of the three vertices, leaving the third out of it.

2.2 Improvement Process

The novel proposal in this work consists of a local search based on the generation of holes (see Algorithm 2) called Piercing Local Search (PLS). This method is proposed with the objective of eliminating vertices from a certain region of the solution. After performing the elimination, the solution will be reconstructed in an intelligent way to obtain better solutions.

The algorithm iterates until it finds no improvement in the entire solution set or reaches the runtime limit (lines 1 to 3). The iteration starts looking for some improvement for a vertex of the dominance set (line 4). To this new solution, the piercing process (line 5) is applied, which will be detailed later in Algorithm 3. Subsequently, the solution is reconstructed using the same greedy criteria described in the constructive procedure (Sect. 2.1) (line 6). Finally, those redundant vertices are removed (line 7) as in Algorithm 1 and, if the obtained solution is better than the previous one (line 8), the best solution is updated (line 9), starting the search again (line 11). Otherwise, it continues searching for an improvement over the next vertex in the set D.

Algorithm 2. *Improve* (D, δ)

```
1:  improve ← TRUE
2:  while improve and time ≤ t_max do
3:      improve ← FALSE
4:      for all v ∈ D do
5:          D' ← Pierce(D, v, δ)
6:          D'' ← Reconstruct(D')
7:          D''' ← RemoveRedundant(D'')
8:          if |D'''| < |D| then
9:              D ← D'''
10:             improve ← TRUE
11:             go to 3
12:         else
13:             D' ← D
14:         end if
15:     end for
16: end while
17: return  D
```

In Algorithm 3 the pseudocode of the piercing process is presented, which aims to pierce a dominance set D' by eliminating several vertices starting from an initial vertex v, avoiding to eliminate those previously fixed by the reduction rules. During this process, all vertices that are δ levels deep from vertex v in the network will be eliminated, the neighborhood with $\delta = 1$ being the set $\{v\}$. In this procedure, if the last expansion level has not been reached (line 1), the vertex v is removed (line 2) and for each of the adjacent vertices (line 3) the algorithm is called with a lower expansion level (line 4), since the level that would correspond to vertex v has been calculated in line 2. Finally, the pierced dominance set D is returned (line 7).

Algorithm 3. *Pierce* (D, v, δ)

```
1:  if δ > 0 then
2:      D ← D \ {v}
3:      for all u ∈ N(v) do
4:          Pierce(D, u, δ − 1)
5:      end for
6:  end if
7:  return  D
```

Starting from the incomplete solutions, the dominant set will be reconstructed by adding the necessary vertices to make it feasible using the greedy degree-based heuristic described in Sect. 2.1.

To graphically illustrate the method, Fig. 1 shows examples of the improvement phase of the implemented BVNS process. The different components of this figure show a social network represented as a graph with 9 vertices and 12 edges at different times of the piercing phase. These vertices are identified with different colors where each color represents a different state. The green ones are those included in the solution set by the constructive method, the black ones are those fixed by the reduction rule of the leaf vertices, and the gray ones by the reduction rule related to the triangles. Meanwhile, the white vertices are not in the solution and the red ones would be excluded by the triangle reduction rule [24]. Finally, orange indicates the initial vertex of the PLS propagation and yellow those affected by the piercing process.

The initial solution proposed by the constructive phase of Fig. 1a has an objective function $|D| = |\{v_0, v_2, v_5, v_6, v_7, v_8\}| = 6$. Applying PLS on vertex v_7, vertices $\{v_1, v_6, v_7, v_8\}$ are affected, which are removed from the solution set (Fig. 1b). It should be noted that vertex v_2 should also be affected, but it is not as it is marked as fixed by the reduction of the leaf vertices. After subsequent greedy reconstruction of the solution set (Fig. 1c), a new solution is obtained such that $|D| = |\{v_0, v_1, v_2, v_5, v_7\}| = 5$, which would be a solution of minimum cardinality for the MPIDS problem.

2.3 Shake Procedure

To enhance the efficiency and exploration of the method of the method, the shaking phase is employed, drawing inspiration from the principles of VNS [18].

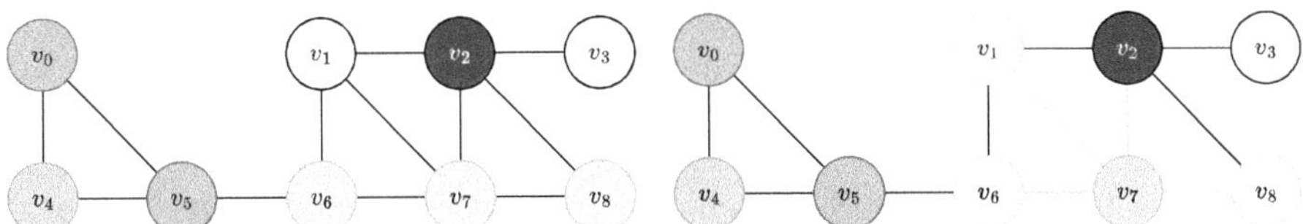

(a) Example of a solution given by an initial construction.

(b) Propagation of the PLS with $\delta = 2$.

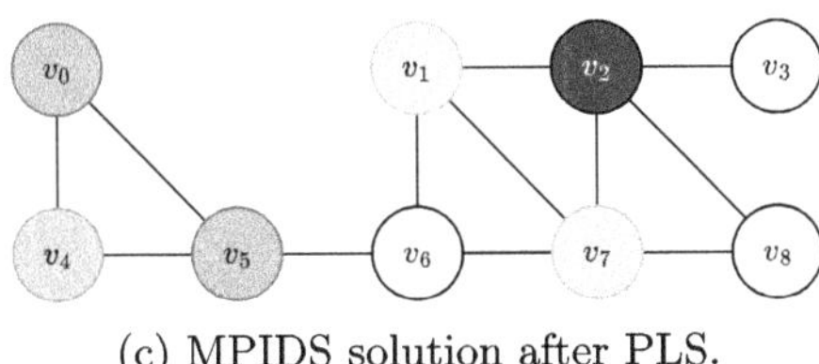

(c) MPIDS solution after PLS.

Fig. 1. Instance composed of 9 vertices and 12 relations, before (a), during (b) and after (c) of the PLS process.

In this context, the shaking phase involves generating a solution D' from a predefined neighborhood $N_k(D)$ of the current solution D. The purpose of this generation is to effectively perturb the solution, thereby enabling the search to escape local optima and explore different regions of the solution space, preventing cycles that could arise from deterministic rules [18]. In this specific implementation, the parameter k defines the number of nodes removed from the current dominating set solution. This process moves the solution to a perturbed state D' from which a new local search can begin. Following this removal, the solution is then intelligently reconstructed.

This reconstructive method will be referred to as the Randomized Greedy Algorithm (RGA). Specifically, the RGA will use the α parameter to determine a subset of vertices, selecting among them the candidate with the highest degree. Here, the α parameter controls the degree of randomness of the method. Specifically, a value of $\alpha = 1$ would denote a fully greedy criterion (since it would evaluate all candidates in the RCL), while a value of $\alpha = 0$ would define a fully random one (since it would only evaluate one candidate at random).

First, we generate a list of vertices NF formed by those vertices that do not yet fulfill the feasibility function such that

$$NF = \left\{ v \in V : |N(v) \cap D| < \left\lceil \frac{|N(v)|}{2} \right\rceil \right\}$$

It is important to note that, for a solution to be feasible, it is necessary to achieve that $NF = \emptyset$, which defines the stopping criteria of the constructive method.

Next, a random element v is chosen from NF, and the candidate list CL that can become part of the solution, formed by $N(v) \setminus D$, is generated, with D being the solution under construction.

Once the CL is available, the restricted candidate list (RCL) is created. While in the traditional scheme this list includes the most promising candidates, in the proposal of this work it uses a fraction α of candidates randomly obtained from the CL, which are the only ones evaluated afterwards. The reason for this modification is that, given the size of the social networks, the evaluation of all candidates from the CL would require a high computational time. The next vertex to be added to the solution will be the one with the highest degree, i.e., $max_{c \in RCL}|N(c)|$. Candidates will continue to be added until the original vertex v meets the feasibility constraints, at which point it will be removed from NF and a new candidate will be chosen.

3   Experimental Results

This section discusses the results obtained in the various experiments carried out. The designed algorithm has been executed on 196 instances collected from the literature, obtained from public repositories such as *Stanford Network Analysis Project* (SNAP) [9] and *Network Repository* [22].

Table 1. Instance characteristics

Metric	Total	Average	Min.	Max.
Size (B)	33 304 329 818	170 791 434	492	4 311 190 746
# Vertices	415 204 269	2 129 253	34	59 216 211
# Edges	2 100 272 636	10 770 628	78	261 321 071

Table 1 lists the main characteristics of these instances. A statistical summary of three key metrics is included: size in bytes, number of vertices and number of edges, along with their total, average, minimum and maximum values. Overall, more than 415 million vertices are considered, where some networks are extremely small (34 vertices), while the largest exceeds 59 million vertices. Meanwhile, more than 2.1 billion edges are recorded, where the smallest network contains 78 connections, while the largest one exceeds 261 million edges.

The designed experimentation consists of two phases and has been performed on a server virtual machine with an AMD EPYC 7282 (2.8 GHz) processor using a single core, 128 GB of RAM, Ubuntu Server 20.04, and Java 21. The first phase includes a preliminary experiment performed on a diverse subset of instances (Sect. 3.1), whose purpose is to define the best parameter values for the algorithm. To avoid overfitting the parameter settings [5], this subset is obtained from the selection of 25% of the available instances. The second phase consists of a final experimentation to validate the best configuration against the state of the art (Sect. 3.2).

All experiments have been evaluated using the same metrics, including: *Average*, which represents the mean of the objective function value of all instances for an algorithm; *Time (s)*, which indicates the average execution time in seconds; *Dev. (%)*, which shows the mean of the percentage deviation of the objective function versus the best known value of the experiment for each instance; and *# Best*, which indicates the number of best solutions found in the experiment. Also, the best of the values in each of these metrics in each experiment has been indicated in bold in each of the tables in this section.

It should be noted that, in the literature, authors establish a stopping criteria of one hour to obtain a solution for some of the instances and 1000 s for others. For this reason, in the experimentation carried out, the execution time of the proposed algorithms has been limited to one hour.

3.1   Preliminary Experiments

Table 2 allows us to analyze the contribution of the different approaches studied: i) a single random construction of the solution (*Random*); ii) multiple random runs limiting the execution time to one hour (*Random (3600 s)*); iii) a single greedy construction (*Greedy*); and iv) a greedy construction with redundant vertex elimination (*Greedy**).

Table 2. Comparison of random and greedy constructions.

Algorithm	Average	Time (s)	Dev. (%)	# Bests
Random	776363.86	3.13	23.23	0
Random (3600 s)	776157.63	3600.00	19.02	3
Greedy	650048.98	**2.74**	20.44	1
*Greedy**	**621659.84**	2.87	**0.42**	**46**

It can be seen how the greedy criteria used is suitable for this problem, since it is able to obtain one of the lowest average values of the objective function and a low deviation in a very short time, compared to the randomized algorithm that uses more than 3000 s. The deviation of this run is particularly noteworthy, because even obtaining a reduction of more than 100000 vertices in the Average metric, it obtains a slightly higher value than the algorithm *Random(3600 s)*. This is due to the large difference in the sizes of the instances, since a difference of 3 or 5 vertices shows between 20% and 30% deviation from the best, while to reach that deviation in large instances there must be a difference of 0.5 million vertices. Since *Greedy* performs better on large instances and worse on small ones, it is penalized with a higher deviation even though it has reduced the Average.

On the other hand, the *Greedy** algorithm with redundant vertices elimination shows even better results than the *Greedy* algorithm. With a slight increase in execution time, it obtains the best value in 46 out of the total 49 instances, leaving the rest of the proposals behind in terms of objective function value and deviation. This shows that the process of eliminating redundant vertices contributes positively in the proposal. Consequently, the *Greedy** method is chosen to validate the δ size of holes within the PLS.

In order to obtain the best configuration of the PLS, two runs have been performed with different values of the δ parameter, collected in Table 3. These values have been 2 and 3, since $\delta = 1$ would only eliminate itself and values of $\delta > 3$ could eliminate the solution almost completely as a consequence of the small world property of networks.

Table 3. Comparison of different values for the δ parameter in the piercing phase.

δ	Average	Time (s)	Dev. (%)	# Bests
1	621659.84	**2.87**	4.25	3
2	565363.71	252.13	**0.13**	40
3	**565325.10**	675.54	0.54	13

With this experimentation, we can observe that, although with a value $\delta = 3$ we obtain a slight improvement in terms of the average value of the objective function, the results with $\delta = 2$ have a smaller deviation and a higher number of better solutions with a notably lower time. This is important since, for large instances, the shorter the time required for local search, the more iterations of BVNS can be performed, allowing better solutions to be found.

In order to find the best configuration of the parameter α in the RGA construct, five runs have been performed with $\delta = 2$ and $\alpha = \{0.00, 0.25, 0.50, 0.75, 1.00\}$ plus a sixth run $\alpha = RND$ where a uniform random value in $\alpha = [0.00, 1.00]$ is generated for each new construct. The execution time considered remains one hour for all configurations except for $\alpha = 0.00$ which, being a greedy construct, always starts from the same initial solution and one execution per instance is required. The results obtained are shown in Table 4.

As can be seen, the value $\alpha = 0.00$ is the one that has obtained the best average, with a shorter time, since its generative criteria is totally greedy. Alternatively, the execution with $\alpha = 1.00$ has obtained a higher quality in terms of better results per number of instances, but obtaining a high average value compared to the rest of the results, which suggests a good performance in small instances, but a weakness with large instances.

Table 4. Comparison for the best configuration of the α parameter in the shake phase for reconstruction.

α	Average	Time (s)	Dev. (%)	# Bests
RND	593147.63	3600.23	0.44	5
0.00	**565363.71**	**252.13**	1.30	6
0.25	615990.47	3600.16	0.51	12
0.50	593060.12	3600.17	0.35	15
0.75	579615.39	3600.22	**0.25**	14
1.00	643084.12	3600.26	0.63	**24**

It has been decided to use the $\alpha = 0.75$ configuration since, in addition to obtaining a competent quality with respect to the two previous α values mentioned above, the best deviation with 0.25% and the second best value in terms of average objective function, it provides a large variability in the construction phase.

As for the k_{max} value, we used $k_{max} = 1$, where 1 is the number of nodes, since higher k_{max} was time-consuming and results in worse results.

3.2   Comparison with the State of the Art

Once the best configuration of the implementation was obtained, a final experimentation has been performed with a greedy initial solution construction, a randomized greedy reconstruction with $\alpha = 0.75$, a redundant vertex elimination, and the PLS with two levels of expansion $\delta = 2$. In order to make a comparison with the best state-of-the-art results obtained by the FastPIDS algorithm designed by Sun et al. [24], the execution time has been limited again to one hour. Note that, compared to the previous study, we have just performed one run instead of using the average of 10 runs. Table 5 collects the results of this comparison.

Table 5. Comparison of the state-of-the-art algorithm and our proposal.

Algorithm	Average	Time (s)	Dev. (%)	# Bests
FastPIDS	**1023870.82**	**3600.00**	**0.01**	**190**
BVNS	1038446.79	3600.03	1.26	18

The evaluation performed reflects that the previous algorithm obtains better results in all the metrics considered, with a lower average value and achieving the best results over 190 instances of the total of 196 instances evaluated. However, the proposed metaheuristic solution obtains reasonable results, with a deviation close to 1.26% and achieving better average in 6 of the instances. The obtained results show how a metaheuristic design based on the combination of several simple methods can show a competitive performance. The next section shows some conclusions and future work described that raise promising ideas for further improvement of these approaches.

4   Conclusion and Future Work

In this work, a BVNS approach has been proposed for the MPIDS problem in social networks. An efficient construction phase has been designed that provides a great variability in the process of obtaining initial solutions on which, in a later improvement stage, it is possible to reduce considerably the size of the dominance set. On the other hand, the proposed PLS generates local optima from the given solutions in a reduced computation time. This constructive and improvement phase, combined with the solution permutation in the BVNS metaheuristic, offers competitive results with respect to FastPIDS.

Some future work derived from this research is to create new intelligent constructive movements [14], as well as studying some alternative greedy criteria to obtain higher quality solutions in a reduced computation time. Finally, we plan to perform a factorial experimentation using tools such as iRACE [15] to obtain the best configuration of the algorithm, selecting the preliminary instances more carefully [16].

Acknowledgments. The authors appreciate the support of the Comunidad Autónoma de Madrid (grant ref. TEC-2024/COM-404), Ministerio de Economía y Competitividad (grant ref. PID2021-125709OA-C22), and Ministerio para la Transformación Digital y de la Función Pública (Cátedra ENIA AI4DDS, grant ref. TSI-100930-2023-3).

References

1. Akbay, M.A., Blum, C.: Application of CMSA to the minimum positive influence dominating set problem. In: Artificial Intelligence Research and Development, pp. 17–26. IOS Press, October 2021. https://doi.org/10.3233/faia210112
2. Akbay, M.A., López Serrano, A., Blum, C.: A self-adaptive variant of CMSA: application to the minimum positive influence dominating set problem. Int. J. Comput. Intell. Syst. **15**(1) (2022). https://doi.org/10.1007/s44196-022-00098-1
3. Bouamama, S., Blum, C.: An improved greedy heuristic for the minimum positive influence dominating set problem in social networks. Algorithms **14**(3), 79 (2021). https://doi.org/10.3390/a14030079
4. Domingos, P., Richardson, M.: Mining the network value of customers. In: Proceedings of the Seventh ACM SIGKDD International Conference on Knowledge Discovery and Data Mining, KDD '01, pp. 57–66. Association for Computing Machinery, New York (2001). https://doi.org/10.1145/502512.502525
5. Heitsch, H., Römisch, W.: Scenario reduction algorithms in stochastic programming. Comput. Optim. Appl. **24**(2), 187–206 (2003). https://doi.org/10.1023/A:1021805924152
6. Herrán, A., Colmenar, J.M., Duarte, A.: A variable neighborhood search approach for the vertex bisection problem. Inf. Sci. **476**, 1–18 (2019). https://doi.org/10.1016/j.ins.2018.09.063
7. King, S.F., Burgess, T.F.: Understanding success and failure in customer relationship management. Ind. Mark. Manage. **37**(4), 421–431 (2008). https://doi.org/10.1016/j.indmarman.2007.02.005
8. Klovdahl, A.S.: Social networks and the spread of infectious diseases: the aids example. Soc. Sci. Med. **21**(11), 1203–1216 (1985). https://doi.org/10.1016/0277-9536(85)90269-2
9. Leskovec, J., Krevl, A.: SNAP Datasets: Stanford large network dataset collection, June 2014. http://snap.stanford.edu/data
10. Lin, G., Guan, J., Feng, H.: An ILP based memetic algorithm for finding minimum positive influence dominating sets in social networks. Physica A Stat. Mech. Appl. **500**, 199–209 (2018). https://doi.org/10.1016/j.physa.2018.02.119
11. Long, C., Wong, R.C.W.: Viral marketing for dedicated customers. Inf. Syst. **46**, 1–23 (2014). https://doi.org/10.1016/j.is.2014.05.003
12. Lozano-Osorio, I., Sánchez-Oro, J., Duarte, A.: An efficient and effective grasp algorithm for the budget influence maximization problem. J. Ambient. Intell. Humaniz. Comput. **15**(3), 2023–2034 (2023). https://doi.org/10.1007/s12652-023-04680-z
13. Lozano-Osorio, I., Sánchez-Oro, J., Duarte, A., Cordón, O.: A quick GRASP-based method for influence maximization in social networks. J. Ambient. Intell. Humaniz. Comput. **14**(4), 3767–3779 (2021). https://doi.org/10.1007/s12652-021-03510-4
14. Lozano-Osorio, I., Sánchez-Oro, J., Sörensen, K.: Determining good solutions and validating them with a metaheuristic approach in social network influence minimization problems. Eur. J. Oper. Res. (2025). https://doi.org/10.1016/j.ejor.2025.11.024
15. López-Ibáñez, M., Dubois-Lacoste, J., Pérez Cáceres, L., Birattari, M., Stützle, T.: The irace package: iterated racing for automatic algorithm configuration. Oper. Res. Perspect. **3**, 43–58 (2016). https://doi.org/10.1016/j.orp.2016.09.002
16. Martín-Santamaría, R., Cavero, S., Herrán, A., Duarte, A., Colmenar, J.M.: A practical methodology for reproducible experimentation: an application to the double-row facility layout problem. Evol. Comput. **32**(1), 69–104 (2024). https://doi.org/10.1162/evco_a_00317

17. Miller, R., Lammas, N.: Social media and its implications for viral marketing. Asia Pacific Pub. Relat. J. **11** (2010)
18. Mladenović, N., Hansen, P.: Variable neighborhood search. Comput. Oper. Res. **24**(11), 1097–1100 (1997). https://doi.org/10.1016/s0305-0548(97)00031-2
19. Pan, J., Bu, T.M.: A fast greedy algorithm for finding minimum positive influence dominating sets in social networks. In: IEEE Conference on Computer Communications Workshops (INFOCOM WKSHPS), IEEE INFOCOM 2019, April 2019, pp. 360–364. IEEE (2019). https://doi.org/10.1109/infcomw.2019.8845129
20. Raei, H., Yazdani, N., Asadpour, M.: A new algorithm for positive influence dominating set in social networks. In: 2012 IEEE/ACM International Conference on Advances in Social Networks Analysis and Mining, August 2012, pp. 253–257. IEEE (2012). https://doi.org/10.1109/asonam.2012.51
21. Robles, J.F., Chica, M., Cordon, O.: Evolutionary multiobjective optimization to target social network influentials in viral marketing. Exp. Syst. Appl. **147**, 113183 (2020). https://doi.org/10.1016/j.eswa.2020.113183
22. Rossi, R.A., Ahmed, N.K.: The network data repository with interactive graph analytics and visualization. In: AAAI (2015). https://networkrepository.com
23. Shan, Y., Kang, Q., Xiao, R., Chen, Y., Kang, Y.: An iterated carousel greedy algorithm for finding minimum positive influence dominating sets in social networks. IEEE Trans. Comput. Soc. Syst. **9**(3), 830–838 (2022). https://doi.org/10.1109/tcss.2021.3096247
24. Sun, R., Wu, J., Jin, C., Wang, Y., Zhou, W., Yin, M.: An efficient local search algorithm for minimum positive influence dominating set problem. Comput. Oper. Res. **154**, 106197 (2023). https://doi.org/10.1016/j.cor.2023.106197
25. Wang, F., Camacho, E., Xu, K.: Positive influence dominating set in online social networks. In: Du, D.-Z., Hu, X., Pardalos, P.M. (eds.) COCOA 2009. LNCS, vol. 5573, pp. 313–321. Springer, Heidelberg (2009). https://doi.org/10.1007/978-3-642-02026-1_29
26. Wang, F., et al.: On positive influence dominating sets in social networks. Theoret. Comput. Sci. **412**(3), 265–269 (2011). https://doi.org/10.1016/j.tcs.2009.10.001

A Machine Learning Enhanced Variable Neighborhood Search Approach for the Uncapacitated Facility Location Problem

Lucas Martín-García[1], Isaac Lozano-Osorio[1(✉)], J. Manuel Colmenar[1], and Belén Melián-Batista[2]

[1] Universidad Rey Juan Carlos, Madrid, Spain
`{lucas.martin,isaac.lozano,josemanuel.colmenar}@urjc.es`
[2] Universidad de La Laguna, La Laguna, Spain
`mbmelian@ull.edu.com`

Abstract. The Uncapacitated Facility Location Problem (UFLP) is widely recognized as a relevant problem in logistics, resource distribution, and telecommunications network planning. Given a set of potential facility locations and a set of customers, the goal is to determine which facilities to open to serve all customers while minimizing both opening and assignment costs. Since this problem is classified as $\mathcal{NP}$-hard, obtaining exact solutions at large scales could not be possible, thereby motivating the use of approximation techniques and metaheuristics. Although early studies used exact formulations derived from the UFLP model, recent research has emphasized the efficacy of approximate and metaheuristic algorithms, which achieve high-quality solutions with substantially reduced computational effort. This work introduces a Variable Neighborhood Search approach to tackle this problem. With the aim of guiding the search toward higher-quality solutions, machine learning techniques have been incorporated to this process. Experimental results on well-known benchmark datasets demonstrate that our method reaches solutions very close to the optimal values, with significantly shorter execution times, outperforming state-of-the-art algorithms validated by the pairwise non-parametric Wilcoxon statistical test.

Keywords: Variable Neighborhood Search · Machine Learning · Uncapacitated Facility Location Problem · Reinforcement Learning

1 Introduction

The Uncapacitated Facility Location Problem (UFLP) is one of the most widely studied optimization problems within the broader family of Facility Location Problems (FLP), devoted to optimally placing facilities to serve customers [5]. Due to its fundamental importance in location theory and its direct applicability to practical domains, ranging from logistics and telecommunications to

S. Cavero et al. (Eds.): ICVNS 2025, LNCS 16256, pp. 15–30, 2026.
https://doi.org/10.1007/978-3-032-19582-1_2

the strategic design of road and railway networks [12], the UFLP has been the subject of extensive research.

Since its introduction [16], the approach to solving the UFLP has evolved significantly. During the latter half of the 20th century, the research was focused on exact algorithms in the UFLP [8] and other FLP variants [1,6], which established a strong formal basis but were limited to instances of moderate size . To overcome this scalability issue, the field later pivoted towards approximate methods, such as the notable greedy local search with progressive cost adaptation [3], designed to handle larger and more complex scenarios. This trend has continued with the development of sophisticated metaheuristics, including genetic algorithms [15], hybrid multistart heuristics [21], and evolutionary algorithms [20], which now dominate the landscape of UFLP solution techniques.

More recently, the best results on this problems were obtained by the Enhanced Group Theory-Based Optimization Algorithm (EGTOA) [23]. As an advanced version of the Group Theory-Based Optimization Algorithm (GTOA) [11], EGTOA has demonstrated superior performance in both solution quality and convergence, outperforming 16 other literature algorithms on the established OR-Library instances. While evolutionary methods like EGTOA are highly effective, this paper explores a different yet powerful paradigm.

A trend in recent years has been the integration of Machine Learning (ML) techniques with metaheuristics to solve complex combinatorial optimization problems [14]. This synergy aims to create a more intelligent search process, improving performance in terms of solution quality, convergence rate, and robustness by extracting useful knowledge from data generated during the search. The literature shows various successful integration points, including enhancing the initialization of solutions and the evolutionary process itself, for instance, through adaptive operator selection.

This approach has proven effective both for the Variable Neighborhood Search (VNS) framework and for problems in the FLP family. For example, Reinforcement Learning has been used within VNS to automate the selection of the most effective neighborhood operators, demonstrating significant improvements in solution quality over conventional VNS implementations [13]. Separately, ML has been applied to the Facility Location Problem by learning from historical data to predict how solutions should be modified in response to parameter perturbations, which can then guide the solver [17].

The demonstrated success of these hybrid strategies provides the primary motivation for this work. In this work, a novel integration of ML into VNS is proposed, specifically designed to guide the constructive phase of the algorithm. Specifically, our approach extracts information of the instance problems, learns from it, and uses the learned information to generate better initial solutions that help the convergence of our algorithmic proposal towards near-optimal solutions for the UFLP. Our experimental results confirm that this VNS-based method yields competitive solutions, establishing it as a viable and effective alternative for tackling the UFLP.

Regarding the structure of the document, the foundational context of the problem and our algorithmic contribution are firstly stated, proceeding then to its empirical validation. More precisely, the paper continues in Sect. 2 with a detailed mathematical formulation of the UFLP and defines the solution representation used in this work. Based on this, Sect. 3 presents our proposed VNS methodology. The effectiveness of the algorithm is then evaluated in Sect. 4, which details the experimental setup and analyzes the results. Finally, the conclusions are drawn in Sect. 5.

2   Problem Formulation

The Uncapacitated Facility Location Problem (UFLP) seeks to determine which subset from a group of potential facilities should be opened to serve a set of customers at a minimum total cost. This total cost is the sum of fixed costs for opening the selected facilities and the costs associated with assigning customers to them. A key feature of the problem is that the number of facilities to be opened is not predetermined.

More formally, let $A = \{a_1, a_2, \ldots, a_M\}$ be a set of M customers and $B = \{b_1, b_2, \ldots, b_N\}$ be a set of N potential facility locations. The cost of serving customer a_i from facility b_j is given by c_{ij} in the cost matrix $C = [c_{ij}]_{M \times N}$. Additionally, each facility b_j has a fixed opening cost d_j.

The problem can be formulated using binary decision variables [15]. For each facility b_j (where $j = 1, \ldots, N$), the variable x_j is equal to 1 if the facility is opened and 0 otherwise. Similarly, the variable y_{ij} is equal to 1 if customer a_i (for $i = 1, \ldots, M$) is assigned to facility b_j, and 0 otherwise. The objective function is presented in Equation (1), while equations (2) through (4) define the constraints that ensure that each customer is assigned to exactly one open facility.

$$\min \quad \sum_{j=1}^{N} d_j x_j + \sum_{i=1}^{M} \sum_{j=1}^{N} c_{ij} y_{ij} \tag{1}$$

$$\text{subject to} \quad \sum_{j=1}^{N} y_{ij} = 1, \quad \forall i \in \{1, 2, ..., M\} \tag{2}$$

$$y_{ij} - x_j \leq 0, \quad \forall j \in \{1, 2, ..., N\}, \forall i \in \{1, 2, ..., M\} \tag{3}$$

$$y_{ij}, x_j \in \{0, 1\}, \quad \forall j \in \{1, 2, ..., N\}, \forall i \in \{1, 2, ..., M\}. \tag{4}$$

A crucial property of the UFLP is that for any given configuration of open facilities, the optimal customer assignment can be determined efficiently. Specifically, each customer is assigned to the open facility that offers the lowest service cost for that customer. Consequently, any solution to the problem can be uniquely and compactly represented by the set of opened facilities, without the need to explicitly state the customer assignments. Therefore, we represent a solution S as the set of open facilities, i.e., $S = \{b_j \in B : x_j = 1\}$. The total cost of a solution S, as determined by the objective function in Eq. (1), will be denoted as $\mathcal{F}(S)$.

To illustrate the difficulty of this problem, consider the example shown in Fig. 1, which consists of four customers (denoted by blue circles) and three potential facilities, b_1, b_2, and b_3 (denoted by empty squares). Each facility has an opening cost of 4 units, and the assignment costs, shown on the edges, correspond to the Euclidean distances.

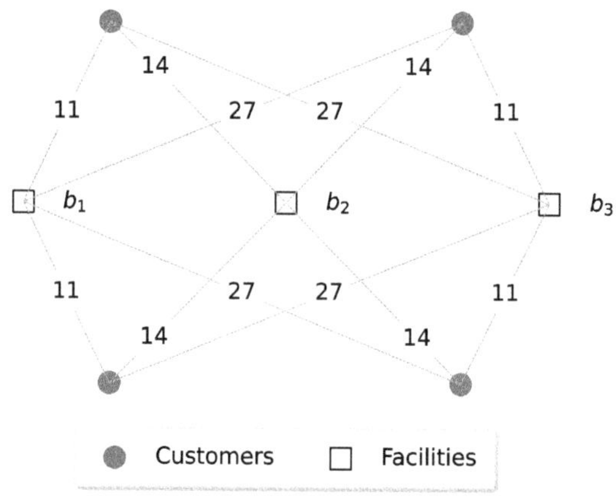

Fig. 1. An instance of the UFLP with four customers (blue circles) and three potential facilities (empty squares). The values on the edges represent the cost of serving each customer from the corresponding facility. (Color figure online)

This simple example highlights the core trade-off of the problem. One possible solution, shown in Fig. 2a, is to open only facility b_2, which is centrally located. This solution, $S_1 = \{b_2\}$, incurs a low opening cost of 4, but the total assignment cost is 56 (14 for each of the four customers), leading to a total cost of $\mathcal{F}(S_1) = 60$. However, a better solution can be obtained by opening two facilities. More precisely, $S_2 = \{b_1, b_3\}$, shown in Fig. 2b. While the opening cost of this solution doubles to 8, the sum of assignment costs is significantly reduced to 44, resulting in a lower total cost of $\mathcal{F}(S_2) = 52$, which is the optimal solution for this instance. This example demonstrates the inherent complexity of the UFLP. Since the number of facilities to open is unconstrained, the number of feasible configurations grows significantly, even for a small instance with only three potential locations.

3   Proposed Algorithm

The Uncapacitated Facility Location Problem (UFLP) is classified as $\mathcal{NP}$-hard [23], which means that exact approaches are often computationally infeasible for

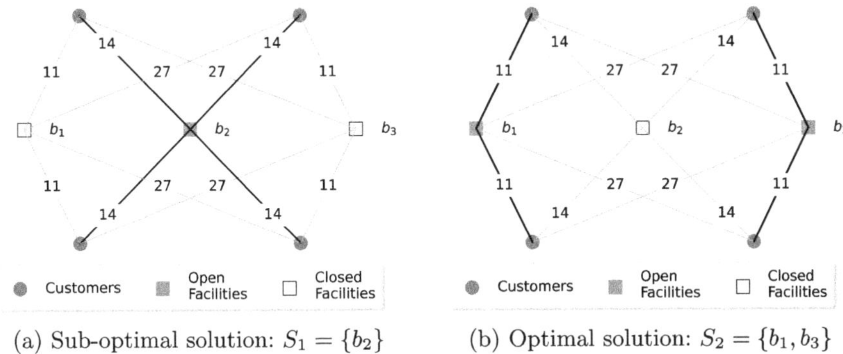

(a) Sub-optimal solution: $S_1 = \{b_2\}$ (b) Optimal solution: $S_2 = \{b_1, b_3\}$

Fig. 2. Alternative solutions for the example instance.

large-scale instances. To address this complexity, this paper proposes an algorithm based on the Variable Neighborhood Search (VNS) metaheuristic, designed to find high-quality solutions within a reasonable timeframe. VNS operates by systematically exploring multiple neighborhood structures, complemented by a perturbation mechanism that allows the search to effectively escape local optima [10,19].

The VNS metaheuristic has been widely and successfully applied to a variety of combinatorial optimization problems. For instance, in the context of location problems, a parallelized VNS has proven highly efficient for the p-Median Problem [9]. Similarly, an optimized version of VNS obtained competitive results for the Capacitated p-Median Problem (CPMP) [7]. More recent applications include a VNS variant for the median problem with interconnected facilities [18] and a parallelized VNS for the Bus Terminal Location Problem (BTLP), which achieved significant improvements in solution quality and computation time [4]. This strong track record motivates its application to the UFLP.

In this work, we adapt VNS for the UFLP to leverage its ability to efficiently explore the solution space and avoid premature convergence. Specifically, our proposal is based on the Basic VNS (BVNS) variant [10]. BVNS effectively balances diversification (through random perturbations) and intensification (through local search), which is key to finding high-quality solutions. The pseudocode for our proposed BVNS algorithm is detailed in Algorithm 1.

The $\texttt{BVNS}(k_{\max})$ method requires the maximum neighborhood size $(k_{\max})$ as an input parameter. The execution begins by generating an initial solution S (line 1) using a constructive method (see Sect. 3.1), which is then set as the best-found solution, denoted S^* (line 2).

The main loop (lines 4–13) iterates until no improvement is found for $k_{\max}$ consecutive neighborhood sizes. In each iteration, the current solution S is perturbed by the $\texttt{Shake}$ function (line 5), which modifies k facilities randomly selected to diversify the search (see Sect. 3.2). Following this, the $\texttt{Improvement}$

Algorithm 1. BVNS($k_{\max}$)

1: $S \leftarrow$ `Constructive()`
2: $S^* \leftarrow S$
3: $k \leftarrow 0$
4: **while** $k < k_{\max}$ **do**
5:     $S \leftarrow$ `Shake`(S^*, k)
6:     $S \leftarrow$ `Improvement`(S)
7:     **if** $\mathcal{F}(S) < \mathcal{F}(S^*)$ **then**
8:         $S^* \leftarrow S$
9:         $k \leftarrow 1$
10:     **else**
11:         $k \leftarrow k + 1$
12:     **end if**
13: **end while**
14: **return** S^*

---

function is applied to the new solution S (line 6) to perform an intensification phase (see Sect. 3.3).

Finally, the algorithm checks if the resulting solution S is better than the best-found solution S^*. If an improvement is found, S^* is updated, and the neighborhood counter k is reset to 1 (lines 7–9). Otherwise, k is incremented (line 11). The size of the perturbation generated by the **Shake** step increases with k. After the termination condition is met, the algorithm returns the best overall solution found, S^* (line 14).

3.1   Constructive Phase

The performance of the BVNS algorithm can be significantly influenced by the quality of its starting point. A well-chosen initial solution can guide the search towards promising regions of the solution space, while a diverse set of starting points can enhance exploration. To this end, we propose and evaluate two different constructive strategies for generating the initial solution S (line 1 of Algorithm 1): a simple random method and a more sophisticated method guided by a reinforcement learning mechanism.

Random Constructive. The first approach is a straightforward random method designed to ensure a valid and diverse starting point with minimal computational cost. The process begins by initializing an empty solution, $S = \emptyset$, which means that all facilities are initially closed. Then, the algorithm iterates through each potential facility and, for each one, decides whether to open it with a probability of 50%.

This stochastic process promotes diversity in the starting points for the VNS. However, it is possible that after iterating through all facilities, the solution set S remains empty. To guarantee a feasible solution where all customers can be served, a post-construction check is performed to ensure at least one facility is open. This simple and computationally inexpensive method ensures a valid starting point and introduces variability, which can help the subsequent search phases to explore different parts of the solution landscape.

Reinforcement Learning Constructive. The second constructive method employs a learning mechanism to intelligently guide the generation of the initial solution. This strategy is designed to obtain information gathered from previous search efforts in the same instance to build progressively better starting solutions over multiple runs of the algorithm. It operates by maintaining a vector of probabilities or rewards, R_k, where each element $R_k(i)$ represents the learned desirability of opening the facility i in iteration k.

Initially, all rewards are set to a neutral value, $R_0(i) = 0.5$ for all facilities $i = 1, \ldots, N$. Consequently, the first execution of this constructive is identical to the random constructive method. A solution S is built by iterating through all facilities and opening each facility i with a probability equal to its current reward value, $R_k(i)$.

The core of this approach lies in its update rule, which is applied after the BVNS algorithm has completed a full execution and returned a final, best-found solution, S^*. The rewards are then updated for the next run of the algorithm based on the composition of S^*. The update rule is defined as follows:

$$R_{k+1}(i) = \min\big(1, \max(0, R_k(i) + \delta(i))\big)$$

where for each facility $i = 1, \ldots, N$:

$$\delta(i) = \begin{cases} +\alpha & \text{if facility } i \in S^* \\ -\alpha & \text{if facility } i \notin S^* \end{cases}$$

Here, α is a learning rate, which we set to $1/N$ to ensure a gradual adaptation. This update mechanism reinforces the selection of facilities that were part of the best-found solution and penalizes those that were not.

This strategy can be framed as an application of Reinforcement Learning by defining its components at a higher level of abstraction. Here, the constructive mechanism itself is the agent, whose policy is parameterized by the reward vector R. A single action consists of generating a complete initial solution S, and a full run of the BVNS algorithm represents one episode. The agent receives a single, delayed reward at the conclusion of the episode, which is implicitly encoded in the final solution S^*. The update rule then adjusts the agent's policy by reinforcing the choices that contributed to this high-quality outcome. By iteratively refining its policy across multiple episodes, the agent learns to generate more effective starting points, concentrating the search effort in more promising areas of the solution landscape.

3.2   Diversification Phase

The **Shake** function, detailed in Algorithm 2, is responsible for the diversification component of the algorithm. Its main purpose is to perturb a given solution S to guide the search to a new neighborhood and escape local optima when the local search can no longer find improvements. The perturbation strategy is based on opening facilities, designed to complement the local search phase, which focuses on closing facilities (see Sect. 3.3). By adding a subset of currently closed facilities to the solution, the search effectively distances itself from the explored region, with the parameter k determining the size of this subset.

Algorithm 2. Shake(S,k)

1: $L \leftarrow \{b_j : b_j \in B \wedge b_j \notin S\}$
2: $L' \leftarrow$ RandomSelection(L,k)
3: $S \leftarrow S \cup L'$
4: **return** S

The process outlined in Algorithm 2 begins by creating an initial list L composed of all facilities that are closed in solution S (line 1). Subsequently, to ensure a varied perturbation, k facilities are selected from this list at random, taking into account that $min(k, |L|)$ (line 2) is the maximum number of facilities that can be opened. Finally, the selected facilities are added to the current solution (line 3), which is then returned by the function (line 4).

3.3   Improvement Phase

The improvement phase of our proposed algorithm serves as the intensification mechanism, where solutions are refined through local search. The core operator for this phase is the closure of an open facility, a move that defines the neighborhoods explored within the BVNS schema. Previous work has shown that this type of move improves both the efficiency and quality of the search [23]. The motivation is that closing facilities reduces the total number of active locations, thereby lowering fixed opening costs and promoting higher-quality solutions. It is also highlighted that maintaining an excessive number of open facilities tends to inflate the objective function value; thus, closing those that do not contribute positively is an effective refinement strategy.

A best improvement strategy has been proposed based on the facility closure move. This procedure evaluates all neighboring solutions, closing the facility that yields the greatest reduction of the objective function.

In addition to the best improvement strategy, a variant based on first improvement was also implemented. In this version, instead of evaluating all neighbors to find the best one, potential facility closures are explored in a random order. The first move that improves the current solution in this random order is immediately accepted. Once an improving closure is found, the solution is updated, and the search process restarts from this new improved solution.

4   Computational Results

This section presents the set of experiments conducted to evaluate the proposed algorithm and compare it against the evolutionary algorithm EGTOA and an exact model, which are considered state-of-the-art references. All experiments were performed on a server equipped with an AMD EPYC 7643 48-core processor capped to 32 cores and 32 GB of RAM. The proposed algorithm was implemented using the Java 21 programming language.

The benchmark instances used in this work are drawn from the literature. Notably, we have expanded upon these by incorporating two additional, widely recognized datasets from related problems. These datasets were evaluated under the same experimental environment to facilitate a robust comparison between algorithms and to analyze their performance across different problem parameters. The characteristics of all instance sets are detailed in Table 1. The first set consists of the Cap instances, which are frequently used for this family of facility location problems and are available from the OR-Library [2][1]. Given the limited size of this initial set, we opted to include more extensive instance sets from the state of the art. Consequently, the K90 and G700 sets [12] were added to the comparison. These instances were generated from the railway and road networks of the Slovak Republic and are openly available[2]. The K90 set originally consisted of 90 individual instances. However, three of them contain negative cost values and were therefore excluded from our analysis to maintain consistency, following the authors of the instances [12]. In summary, a total number of 802 instances were studied, ranging from 16 to 1000 facilities and 50 to 2906 customers.

Table 1. Description of the benchmark instance sets studied.

Name	Instances	Facilities	Customers	Reference
Cap	15	16–100	50–1000	[2]
K90	87	45–457	457	[12]
G700	700	100–1000	2906	[12]

The EGTOA algorithm was executed on the new instance sets, which was made possible by the authors' decision to publish the source code alongside their article [23], allowing for its adaptation and application in this study. The exact model, on the other hand, was implemented in Gurobi using the mathematical formulation of the problem.

The results tables in our experiments present four key metrics to evaluate algorithm performance. The metric denoted as $\mathcal{F}$ reflects the average objective function value obtained by each algorithm, where a lower value indicates a better

---

[1] https://people.brunel.ac.uk/~mastjjb/jeb/orlib/uncapinfo.html.
[2] http://frdsa.uniza.sk/~buzna/supplement.

solution. *Time (s)* represents the average computation time in seconds required for each algorithm to run. *Gap (%)* measures the average percentage difference between the solutions found by an algorithm and the optimal solutions, providing a measure of relative error. Finally, *# Opt* indicates the number of instances for which each algorithm found the optimal solution.

To ensure the robustness of our results and account for the stochastic elements of the algorithm, the proposed VNS variants, the random constructive based BVNS-R and the reinforcement learning constructive based BVNS-RL, were executed 30 times each for every instance. This multi-run approach serves a dual purpose. For the BVNS-R variant, it ensures that the search begins from multiple, diverse random starting points. For the BVNS-RL variant, it provides the necessary number of episodes for the reinforcement learning mechanism to effectively learn and adapt its constructive strategy based on previous outcomes in the same instance. The results presented in the tables are aggregated from these 30 runs as follows: the objective function ($\mathcal{F}$) and Gap (%) values reflect the best solution found across all 30 executions, while the reported Time (s) represents the total cumulative time for the full 30 runs to complete.

The experiments are structured in two stages: a preliminary phase and a final comparison phase. The preliminary phase involves conducting tests to determine the best parameter values for each variant of the proposed algorithm. Subsequently, the final phase evaluates the performance of the selected best configurations, comparing them against the other aforementioned methods.

4.1 Preliminary Results

The preliminary experimental phase is designed to tune the parameters of our proposed algorithm. This was carried out on a representative subset of the instances, created by randomly selecting 20% of the instances from each dataset using stratified sampling [22]. This approach ensures that the preliminary subset maintains the proportional representation and diversity of the full benchmark, which includes instances with varying origins, sizes, and cost distributions.

The primary goal of this phase is to determine the optimal configuration of the algorithms considering two variants depending on the constructive method: Random (BVNS-R), and Reinforcement Learning (BVNS-RL). The configuration implies determining the local search strategy (First Improvement vs. Best Improvement) and the perturbation size factor, K, for which we tested the values $\{0.1, 0.2, 0.3, 0.4, 0.5\}$. The parameter K is used to define the maximum neighborhood size in Algorithm 1 as $k_{\max} = \lceil N \cdot K \rceil$, where N is the total number of potential facilities. The results for each of the two variants are presented and analyzed separately.

Table 2 shows the aggregated results for the BVNS-R variant. A clear observation across all values tested for K is that the Best Improvement local search strategy consistently and significantly outperforms the First Improvement strategy across all metrics: it achieves lower objective function values ($\mathcal{F}$), substantially smaller Gap values, finds a much higher number of optimal solutions (# Opt) and spends a shorter computational time.

Focusing on the Best Improvement strategy to select the best value for K, we analyze the trade-off between the quality of the solution and the computational time. As expected, execution time is directly proportional to K, as a larger maximum perturbation size increases the computational effort of the local search. While $K = 0.1$ is the fastest configuration, increasing the parameter to $K = 0.2$ yields a dramatic improvement in solution quality reducing the Gap by more than 70% (from 0.006122% to 0.001724%). This substantial gain justifies more than doubling the execution time. However, a further increase to $K = 0.3$ nearly doubles the time again, but offers only a marginal improvement in the Gap (from 0.001724% to 0.001707%). This indicates diminishing returns for larger values of K. Therefore, we select $K = 0.2$ as the best-balanced configuration for the Random Constructive variant BVNS-R, as it provides an excellent compromise between computational cost and solution quality.

Table 2. Preliminary results for the BVNS algorithm with the Random Constructive, BVNS-R, evaluated on 20% of the dataset.

Local Search	K	$\mathcal{F}$	Time (s)	Gap (%)	# Opt
	0.1	$3,174,785.33$	792.34	0.274205	58
	0.2	$3,170,779.44$	1007.23	0.233152	66
First Improvement	0.3	$3,171,017.58$	1240.25	0.243865	65
	0.4	$3,171,210.25$	1502.95	0.244462	66
	0.5	$3,170,757.61$	1728.05	0.233776	65
	0.1	$3,167,364.00$	**93.92**	0.006122	133
	0.2	$3,167,281.28$	227.00	0.001724	148
Best Improvement	0.3	$3,167,279.87$	423.14	0.001707	**151**
	0.4	$3,167,278.90$	684.37	0.001595	**151**
	0.5	$\mathbf{3,167,276.27}$	1017.10	**0.001489**	150

The results for the BVNS-RL variant are presented in Table 3. As with the random version, the Best Improvement local search is unequivocally superior to First Improvement, yielding better results in all metrics for all tested K values.

Therefore, we proceed to select the optimal K value for the Best Improvement strategy. The analysis reveals a similar pattern. Moving from $K = 0.1$ to $K = 0.2$ leads to a significant improvement in solution quality, with the Gap dropping from 0.210642% to 0.001082%. This significant leap in performance strongly justifies the increase in execution time from 88 s to 225 s. In contrast, increasing the parameter further to $K = 0.3$ almost doubles the runtime again, but provides a much smaller relative improvement in the Gap (to 0.000836%). This again points to $K = 0.2$ as the best value in this balance. Although other configurations achieve better individual metrics (e.g., $K = 0.3$ has the best $\mathcal{F}$ and Gap), the configuration with $K = 0.2$ provides a solution of nearly identical quality with less computational cost.

Table 3. Preliminary results for the BVNS algorithm with the Reinforcement Learning Constructive, BVNS-RL, evaluated on 20% of the dataset.

Local Search	K	$\mathcal{F}$	Time (s)	Gap (%)	# Opt
First Improvement	0.1	$3,177,592.01$	939.49	0.497938	58
	0.2	$3,172,158.29$	1093.90	0.245668	65
	0.3	$3,171,408.92$	1316.49	0.239123	66
	0.4	$3,174,379.58$	1598.42	0.262672	63
	0.5	$3,170,757.10$	1853.15	0.232076	63
Best Improvement	0.1	$3,169,410.98$	**88.01**	0.210642	139
	0.2	$3,167,271.68$	224.93	0.001082	**151**
	0.3	$\mathbf{3,167,269.01}$	424.91	**0.000836**	**151**
	0.4	$3,167,276.50$	683.34	0.001309	147
	0.5	$3,167,271.70$	1020.71	0.001015	149

Based on this preliminary analysis, we conclude that the Best Improvement is the most suitable local search strategy for this problem. For the perturbation size, $K = 0.2$ provides the best overall trade-off between execution time and solution quality for both constructive methods. Therefore, the configurations selected for the final comparison are BVNS-R (Random Constructive) and BVNS-RL (RL Constructive), both using the Best Improvement local search and a $K = 0.2$ value.

4.2   Final Results

Following the parameter tuning in the preliminary phase, the two best configurations of our proposed algorithm, BVNS-R and BVNS-RL, were evaluated against the state-of-the-art EGTOA algorithm and the exact model. This final comparison was performed on the entire set of benchmark instances. The comprehensive results, broken down by instance set as well as aggregated, are presented in Table 4.

On the standard Cap instances, both the EGTOA and BVNS-RL algorithms successfully find all 15 optimal solutions, matching the performance of the Exact Model. The BVNS-R variant performs nearly as well, finding 14 optima with a low Gap of 0.00218%. In terms of computational time, our proposed VNS methods are exceptionally efficient, with BVNS-RL being the fastest overall at 0.208 s, significantly outperforming both the exact model (0.802 s) and the much slower EGTOA (258.93 s).

For the K90 instance set, the performance differences become more pronounced. EGTOA struggles significantly, resulting in a large average Gap of 12.13% and failing to find any optimal solutions, all while requiring over 1,400 s. In contrast, both BVNS-R and BVNS-RL achieve near-optimal results, with gaps of just 0.00568% and 0.00374%, respectively. Their execution times are

Table 4. Final results comparing EGTOA, the proposed BVNS variants, and the exact model.

Instances	Algorithm	$\mathcal{F}$	Time (s)	Gap (%)	# Opt
Cap	EGTOA	$3,502,545.59$	258.931	0.00000	15
	BVNS-R	$3,502,796.74$	0.254	0.00218	14
	BVNS-RL	$3,502,545.59$	0.208	0.00000	15
	Exact Model	$3,502,545.59$	0.802	0.00000	15
K90	EGTOA	$3,069,045.61$	$1,405.238$	12.12858	0
	BVNS-R	$2,765,428.69$	3.825	0.00568	71
	BVNS-RL	$2,765,390.11$	3.508	0.00374	73
	Exact Model	$2,765,298.57$	2.642	0.00000	87
G700	EGTOA	$27,202,096.73$	$3,597.980$	174.81629	23
	BVNS-R	$2,943,809.46$	248.588	0.00018	666
	BVNS-RL	$2,943,809.24$	239.329	0.00010	678
	Exact Model	$2,943,809.00$	62.861	0.00000	700
Aggregated	EGTOA	$24,140,913.79$	$1,236.663$	153.89849	38
	BVNS-R	$2,934,913.80$	217.392	0.00081	751
	BVNS-RL	$2,934,904.73$	209.275	0.00050	766
	Exact Model	**2,934,894.58**	**55.168**	**0.00000**	**802**

also highly competitive, remaining in the same order of magnitude as the exact model and finding a large majority of the optimal solutions.

The trend continues and is further amplified on the large-scale G700 instances. Here, EGTOA's performance degrades severely, yielding a very high Gap of 174.8% and requiring nearly an hour of computation time. Our VNS variants, however, maintain their excellent performance, delivering solutions with gaps close to zero and finding almost all 700 optimal solutions. While the exact model is fastest on this set, our methods provide a robust and effective heuristic alternative, demonstrating strong scalability.

The aggregated results provide a clear overview of overall performance. The state-of-the-art EGTOA is shown to be non-competitive on the broader set of instances, with an extremely high average gap and long execution time. In contrast, both of our proposed methods, BVNS-R and BVNS-RL, deliver solutions of exceptional quality, with average gaps of just 0.00081% and 0.00050%, respectively. While the exact model is the fastest on average, our algorithms are orders of magnitude faster than EGTOA. Furthermore, when comparing our two proposed variants, the BVNS-RL consistently shows a slight edge over BVNS-R across all aggregated metrics: a better objective function value, a smaller gap, more optima found, and a faster runtime. This confirms the positive contribution of the Reinforcement Learning mechanism. To conclude the experimental section, our VNS-based approaches, particularly BVNS-RL, prove to be highly effective and efficient alternatives for solving the UFLP, providing near-optimal results with remarkable consistency and speed.

Finally, we apply the widely used non-parametric Wilcoxon signed-rank test for pairwise comparisons to investigate whether the solutions generated by the two algorithms originate from distinct populations. The obtained p-value, which is smaller than 0.0001 when comparing BVNS-RL with EGTOA and BVNS-R with EGTOA, provides strong evidence supporting the superior performance of the proposed BVNS-RL and BVNS-R algorithms.

5   Conclusions

In this paper, we have proposed and evaluated a Variable Neighborhood Search algorithm for the Uncapacitated Facility Location Problem. The methodology is built upon a Basic VNS framework that strikes a balance between intensification and diversification, using an improvement phase based on closing facilities under two different strategies, and a perturbation mechanism that opens new ones. We explored two distinct constructive heuristics to generate the initial solutions: a simple random approach and a more sophisticated method guided by a Reinforcement Learning mechanism.

The computational results analyzed a total number of 802 instances from well-known benchmarks including the large-scale K90 and G700 sets. The experiments demonstrated the effectiveness and efficiency of our proposed VNS approach. Both variants of our algorithm deliver highly competitive solutions, achieving near-zero deviation from the verified optimal solutions. In comparison to the state-of-the-art EGTOA algorithm, our methods show vastly superior performance, particularly on larger and more complex instances where EGTOA's solution quality degraded significantly. The analysis also revealed that the Reinforcement Learning constructive (BVNS-RL) consistently provided a slight but clear improvement over the random constructive (BVNS-R) in all performance metrics, highlighting the value of incorporating a learning component to guide the search towards promising regions of the solution space.

For future work, several research directions are identified. One practical avenue is to enhance computational efficiency by investigating methods to accelerate the objective function calculation, which is a bottleneck during the intensive local search phase. Another promising direction is to further explore the integration of different Machine Learning techniques. This could involve not only developing more advanced constructive heuristics but also applying Machine Learning to other parts of the algorithm, such as the dynamic selection of neighborhoods or guiding the perturbation strategy, thereby further enhancing the algorithm's performance and adaptability.

Acknowledgment. This work was made possible through the support of the Comunidad Autónoma de Madrid (grant ref. TEC-2024/COM-404), Ministerio de Economía y Competitividad (grants ref. PID2021-125709OA-C22, PID2019-104410RB-I00/AEI/ 10.13039/501100011033), and Ministerio para la Transformación Digital y de la Función Pública (Cátedra ENIA AI4DDS, grant ref. TSI-100930-2023-3).

References

1. Akinc, U., Khumawala, B.M.: An efficient branch and bound algorithm for the capacitated warehouse location problem. Manage. Sci. **23**(6), 585–594 (1977). https://doi.org/10.1287/mnsc.23.6.585
2. Beasley, J.E.: Or-library: distributing test problems by electronic mail. J. Oper. Res. Soc. **41**(11), 1069–1072 (1990). http://www.jstor.org/stable/2582903
3. Charikar, M., Guha, S.: Improved combinatorial algorithms for the facility location and k-median problems. In: 40th Annual Symposium on Foundations of Computer Science (Cat. No. 99CB37039), pp. 378–388 (1999). https://doi.org/10.1109/SFFCS.1999.814609
4. Djenić, A., Radojičić, N., Marić, M., Mladenović, M.: Parallel VNS for bus terminal location problem. Appl. Soft Comput. **42**, 448–458 (2016). https://doi.org/10.1016/j.asoc.2016.02.002
5. Drezner, Z., Hamacher, H.: Facility Location: Applications and Theory. Springer, Heidelberg (2002)
6. Efroymson, M.A., Ray, T.L.: A branch-bound algorithm for plant location. Oper. Res. **14**(3), 361–368 (1966). https://doi.org/10.1287/opre.14.3.361
7. Fleszar, K., Hindi, K.: An effective VNS for the capacitated p-median problem. Eur. J. Oper. Res. **191**(3), 612–622 (2008). https://doi.org/10.1016/j.ejor.2006.12.055
8. Galvão, R.D., Raggi, L.A.: A method for solving to optimality uncapacitated location problems. Ann. Oper. Res. **18**(1), 225–244 (1989). https://doi.org/10.1007/bf02097805
9. García-López, F., Melián-Batista, B., Moreno-Pérez, J.A., Moreno-Vega, J.M.: The parallel variable neighborhood search for the p-median problem. J. Heuristics **8**(3), 375–388 (2002). https://doi.org/10.1023/a:1015013919497
10. Hansen, P., Mladenović, N.: Variable neighborhood search: principles and applications. Eur. J. Oper. Res. **130**(3), 449–467 (2001). https://doi.org/10.1016/S0377-2217(00)00100-4
11. He, Y., Wang, X.: Group theory-based optimization algorithm for solving knapsack problems. Knowl. Based Syst. **219**, 104445 (2021). https://doi.org/10.1016/j.knosys.2018.07.045
12. Janáček, J., Buzna, Ľ: An acceleration of erlenkotter-körkel's algorithms for the uncapacitated facility location problem. Ann. Oper. Res. **164**(1), 97–109 (2008). https://doi.org/10.1007/s10479-008-0343-0
13. Kalatzantonakis, P., Sifaleras, A., Samaras, N.: A reinforcement learning-Variable neighborhood search method for the capacitated Vehicle Routing Problem. Exp. Syst. Appl. **213**, 118812 (2023). https://doi.org/10.1016/j.eswa.2022.118812
14. Karimi-Mamaghan, M., Mohammadi, M., Meyer, P., Karimi-Mamaghan, A.M., Talbi, E.G.: Machine learning at the service of meta-heuristics for solving combinatorial optimization problems: a state-of-the-art. Eur. J. Oper. Res. **296**(2), 393–422 (2022). https://doi.org/10.1016/j.ejor.2021.04.032
15. Kiran, M.S.: The continuous artificial bee colony algorithm for binary optimization. Appl. Soft Comput. **33**, 15–23 (2015). https://doi.org/10.1016/j.asoc.2015.04.007
16. Kuehn, A., Hamburger, M.: A heuristic program for locating warehouses. Manage. Sci. **9**(4), 643–666 (1963). https://doi.org/10.1287/mnsc.9.4.643
17. Lodi, A., Mossina, L., Rachelson, E.: Learning to handle parameter perturbations in Combinatorial Optimization: an application to facility location. EURO J. Transp. Logist. **9**(4), 100023 (2020). https://doi.org/10.1016/j.ejtl.2020.100023

18. Lozano-Osorio, I., Sánchez-Oro, J., López-Sánchez, A.D., Duarte, A.: A variable neighborhood search for the median location problem with interconnected facilities. Int. Trans. Oper. Res. **32**(1), 69–89 (2025). https://doi.org/10.1111/itor.13468
19. Mladenović, N., Hansen, P.: Variable neighborhood search. Comput. Oper. Res. **24**(11), 1097–1100 (1997). https://doi.org/10.1016/s0305-0548(97)00031-2
20. Rahkar Farshi, T.A., Agahian, S., Dehkharghani, R.: BinBRO: binary battle royale optimizer algorithm. Exp. Syst. Appl. **195**, 116599 (2022). https://doi.org/10.1016/j.eswa.2022.116599
21. Resende, M.G., Werneck, R.F.: A hybrid multistart heuristic for the uncapacitated facility location problem. Eur. J. Oper. Res. **174**(1), 54–68. https://doi.org/10.1016/j.ejor.2005.02.046
22. Singh, R., Mangat, N.S.: Stratified sampling. In: Singh, R., Mangat, N.S. (eds.) Elements of Survey Sampling, pp. 102–144. Springer, Dordrecht (1996). https://doi.org/10.1007/978-94-017-1404-4_5
23. Zhang, F., He, Y., Ouyang, H., Li, W.: A fast and efficient discrete evolutionary algorithm for the uncapacitated facility location problem. Exp. Syst. Appl. **213**, 118978 (2023). https://doi.org/10.1016/j.eswa.2022.118978

Adaptive VNS in Dynamic Optimization of Touristic Itineraries

Cristina González-Navasa[1,2], Helí Alonso Afonso[2,3],
José A. Moreno Pérez[3,4(✉)], and Julio Brito Santana[3,4]

[1] Dep. of Data Science, Atlantis Technology, 38206 La Laguna, Spain
cgnavasa@atlantistecnologia.com
[2] University of La Laguna, 38271 La Laguna, Spain
halonsoa@ull.edu.es
[3] Dep. Informática y Sistemas, University of La Laguna, 38271 La Laguna, Spain
{jamoreno,jbrito}@ull.edu.es
[4] Instituto Universitario de Desarrollo Regional, University of La Laguna,
La Laguna, Spain

Abstract. We propose an Adaptive Variable Neighbourhood Search
(AVNS) metaheuristic for dynamic optimisation of tourist itineraries at
destination for use in an interactive online tourist route recommender.
The recommendation must use up-to-date information and be able to
adapt the recommendation according to changes in the circumstances of
the touristic activities or to specific requirements of the tourist user.
Instead of using an algorithm to get the new recommendation from
scratch after the changes, we propose to apply AVNS from the previous
proposal. The adaptive mechanism takes into account previous performance to tune the parameters to the change that occurred. We test the
proposal using real data from the tourist island of Tenerife in the Canary
Islands. The results show that our proposal provides better results in less
computational time.

Keywords: Touristic Itineraries · Adaptive VNS · Machine Learning ·
GRASP

1 Introduction

The purpose of this work is to explore the use of an *Adaptive Variable Neighbourhood Search* (AVNS) metaheuristic for dynamic route optimization to include it
in an *interactive online tourist itinerary recommender*. Tourism is among the
most dynamic and influential sectors globally and is the main driver of the
economy of a growing number of regions and countries. It is an activity with
rapid and constant evolution driven by technological advances and the increasing
connectivity and availability of data. The digital transformation conditions the
tourist behaviour, facilitating real-time decision making for personalized experiences with the help of recommendation systems. These systems integrate data
update and advanced algorithms to help travellers better plan their itineraries at

S. Cavero et al. (Eds.): ICVNS 2025, LNCS 16256, pp. 31–45, 2026.
https://doi.org/10.1007/978-3-032-19582-1_3

destinations. These recommenders should select points of touristic interest and optimize itineraries using updated data to take into account dynamic constraints and personal preferences [45,53].

This type of recommender must use up-to-date information about the tourist user and the circumstances of the possible points of interest to visit in order to make the most appropriate recommendation in each case. The development of an interactive online tourist route recommender requires a tool that can reformulate the proposal, totally or partially, in the event of unforeseen events, changes in circumstances that affect the touristic activity, or at the request of the tourist user. The optimization problems associated with tourist itinerary recommendations are known as *Tourist Trip Design Problems* (TTDP) [15]. The main parameters and data for them are the scores associated with the *Points of Interest* (POIs) that can be visited, corresponding to the satisfaction of the tourist by visiting them, and the time used to visit them and travel between the points. The objective is to maximize the joint score of POIs in the itinerary, subject to the corresponding constraints. The formulation models for these problems derive from the combinatorial models of the *Orienteering Problems* (OP) that seek to select the route that visits a subset of points that maximizes the total cumulative score without exceeding a given time limit, taking into account the visit times and the travel times [21,63]. For a multi-day stay, the models correspond to the *Team Orienteering Problem* (TOP) where a given set of routes (one for each day of the stay and starting and ending at the tourist hotel or accommodation) with mutually exclusive POIs has to maximize the total score [7]. In realistic cases, these scores and times vary depending on several aspects [49].

In addition to the usual constraints in routing problems that guarantee the feasibility of the routes, additional constraints are used to ensure the limited duration for each route. Some versions of TTDP include *Time Windows* (TW) to reflect the opening hours of the POIs. Other extensions allow for the limitation of the number of thematic groupings of points of interest (POIs) (such as beaches), mandatory visits (such as certain monuments), or the necessity of including a restaurant each day. Models often incorporate multiple constraints [16,22,29,35, 36,48,54,56]. Other extensions, such as *time-dependent* versions or additional variants, have also been addressed [33,34,64].

The *Variable Neighbourhood Search* (VNS) is a metaheuristic that has been shown to be efficient for a large number of route optimization problems, including several derived from the Team Orienteering Problem (TOP) [3,25]. However, dealing with the various optimization problems associated with the redesign of the recommended touristic itineraries in the face of the large number of circumstances that may motivate the need to modify the proposal requires a metaheuristic that adapts to such changes. The effective application of VNS requires the tuning of a series of neighbourhoods associated with specific movements in the search space to be applied to an initial proposal. Adaptive variants of the VNS incorporate some component that dynamically performs the selection and tuning of the neighbourhoods that are applied to the proposed solution within the structure of the VNS. These components try to react to the performance of each of the possible moves in order to select the most appropriate at each stage of the process. In this work, an Adaptive VNS is implemented and experimentally analyzed on a benchmark of instances constructed from real data of points of touristic interest on the island of Tenerife. GRASP (Greedy Randomized Adaptive Search Procedure) [13,44] is, with VNS, the metaheuristic that has been most often applied to TTDP [49]. Our adaptive procedure is derived from a hybrid between VNS and GRASP that has already been applied to TTDP [19].

The remainder of the paper is organized as follows. The next section includes a short literature review on metaheuristics applied to TTDP and Adaptive VNS in routing. Section 3 includes the basics of Adaptive VNS. Section 4 describes the details of the hybrid metaheuristic used to solve this problem. The paper ends with the experimentation section and the conclusions.

2   Literature Review

We include a limited review of published approaches for the Tourist Trip Design Problem (TTDP), focused on metaheuristic methods, and existing adaptive versions of Variable Neighbourhood Search (VNS), focused on those applied to routing problems. A more comprehensive review is not possible due to the limitation in space.

TTDP and most of its variants are NP-hard problems [21]. Therefore, instead of exact resolution, heuristics and metaheuristics are commonly used. A long series of metaheuristics and their extensions and hybrids have been tested for these problems. According to the systematic review of the literature on TTDP by Ruiz-Meza and Montoya-Torres in 2022 [49], the most used metaheuristics to deal with TTDP are Iterated Local Search, [55], Simulated Annealing, [1,10], Genetic Algorithm [14,18], GRASP (Greedy Randomized Adaptive Search Procedure) [13,44] and VNS (Variable Neighbourhood Search) [25,41].

The systematic review of the literature on TTDP solution techniques [49] includes papers from the period 2010–2021. So we here briefly review the papers that use metaheuristics to address some version of the TTDP during the period 2022–2025. A Simulated Annealing algorithm has been applied in [40]; [66] considers a hybrid approach from Genetic Algorithms and Differential Evolution, and [17] combines a Genetic Algorithm with VNS, namely with a VNS version known as *Variable Neighbourhood Descent* (VND). In addition, [68] uses a greedy type algorithm, and [11] applied a hybrid approach based on VNS with Mathematical Programming. GRASP is used in [52] and [20] and VNS in [33] and [39]. Hybrids that combine GRASP and VNS are applied in [47] and [19]. As shown in the review mentioned above, GRASP and VNS are the metaheuristics most commonly used applied to TTDP until 2022. These two metaheuristics have also often been jointly applied to other routing problems, for example, in 2025 we find [46] and [9]. However, those hybridizations between GRASP and VNS basically consist of using VNS in the GRASP optimization phase. We proposed a different way to hybridize these two well known metaheuristics [19]. One of the first applications of GRASP to a TTDP appears in [63]. On the other hand, one of the first applications of VNS to TTDP dates back to 2007 [2] and several of the main versions of TTDP have also been solved with VNS by 2009 [61,62].

The literature on adaptive VNS is too large to fit in the size of this paper. We briefly review only the literature related to route optimization. One of the first proposals of AVNS was an adaptive VND applied to a multidepot vehicle routing problem with time windows [43]. Adaptive VND was also applied for collection routing in [12] An adaptive BVNS is applied to home health care routing [6]. AVNS was applied to solve the multi-compartment vehicle routing problem in [67] and [4]. A double-adaptive general VNS has been used for an unmanned electric vehicle routing problem in [38] VNS with reinforcement learning is used in [8,51], and [30] to deal with the Traveling Salesman problem, the periodic Vehicle Routing problem, and the Capacitated Vehicle Routing problem, respectively. An hybridization of VNS and Large Neighbourhood Search for a Location-Routing Problem is applied in [60]. Moreover, [32] and [69] use adaptivity to address a TTDP but not with VNS. Finally, we highlight the use of AVNS for a dynamic VRP in [58] where interruptions in communication routes between the points to be visited are considered.

3   Adaptive VNS

Artificial Intelligence (AI) and *Machine Learning* (ML) are nowadays present in all scientific and technological fields, particularly tourism [23,24,50,53] and optimization [31,59], where they are increasingly being used to improve the performance of proposals. *Adaptive VNS* (AVNS) is the result of applying these ideas to the VNS metaheuristic versions. Variable Neighbourhood Search (VNS) [25] is a metaheuristic originally proposed by Nenad Mladenović in [42], whose key idea is to apply systematic changes in the structure of neighbourhoods during the search. VNS plays an important role in the field of metaheuristics and their applications in many areas of engineering and optimization [3,5,25,27,28]. The key purpose is to take advantage of a change in the movements considered when performing a neighbourhood search. The first articles presenting the VNS metaheuristic, [41] and [26], already highlighted the importance of managing the neighbourhood structures involved in a VNS. The following paragraph appears there:

> "When using more than one neighbourhood structure (in the search),
> the following problem-specific questions have to be answered:
> 1. What should be used and how many of them?
> 2. What should be their order in the search?
> 3. What strategy should be used in changing neighbourhoods?"

Adaptive VNS algorithms arise from the use of AI techniques and ML tools to provide adaptive responses to these issues. Until the arrival of these advances, the general recommendation of specialists was to use neighbourhood structures arranged in order of size or complexity, from the simplest and fastest to the most complex and sophisticated. Different adaptive strategies used with metaheuristics ([31]) are embodied in the method of assigning credits to the possible operators, the *Credit Assignment* method, and in the selection mechanism between such operators, the *Selection Assignment* method. We use the simplest strategy; a *Score Credit Assignment* (SCA) where the scores of the neighbourhoods or operators are updated based on their performance in the process and a *Roulette-Wheel Selection* (RWS) which assigns selection probabilities proportional to the credit of each value, so that values with higher credits have a higher probability of being selected. RWS is a classical selection rule [37] often used in Simulation and in Genetic Algorithms since its foundation [18]. With this method, the probability of selection is proportional to the operator score. It can be briefly described as follows. Consider a series of N operators $\{o_1, o_2, \ldots, o_N\}$, each with its score s_i and let $S = s_1 + s_2 + \cdots + s_N$. The selection probability of the i-th operator is thus given by $p_i = s_i/S$ for $i = 1, 2, \ldots, N$. Let us imagine a roulette wheel with sectors of size proportional to s_i. The selection of an operator is equivalent to randomly choosing a point on the wheel and locating the corresponding sector. In a roulette-wheel selection, one constructs a line segment of length s_i out of consecutive sectors of total length S, generates a random number r in the interval $[0, W]$, and locates the corresponding sector, thus selecting the respective operator o_i.

4   Adaptive VNS-GRASP Hybrid

GRASP is a very often metaheuristic used in itinerary optimization due to its performance and the similarity with the natural behavior of the tourists. An intuitive greedy heuristic traditionally used by tourists for choosing their itinerary at destinations consists of selecting, one by one, POIs guided by some greedy functions that include scores and travel times. We use the greedy *effectiveness* function of every new POI that, given an itinerary, is the quotient between the score of this POI and the increment in the duration of the itinerary due to its inclusion in the itinerary. Every new POI included in an itinerary is always inserted in its best position; that with minimum increment in the duration of the itinerary. We use this function in the constructive phase of GRASP where, instead of iteratively selecting the best new POI to be included in the solution according to its effectiveness, it is randomly selected from a *Restricted Candidate List* (RCL). Given a limit size L of RCL, at each iteration it consists of the L best POIs to insert according to their effectiveness and taking into account the constraints. The constructive phase iteratively selects a new POI from RCL and ends when the RCL is empty.

The classic GRASP metaheuristic iteratively applies the constructive phase (*the constructive phase*) followed by some local search or improvement phase (*the improvement phase*). The traditional GRASP-VNS hybrids consist of using a version of VNS in the improvement phase of GRASP; usually a VND. The hybrid metaheuristic we propose combines VNS and GRASP but in a different way. Instead of only using VNS as the GRASP improvement phase, we include the GRASP constructive procedure inside VNS [19]. This hybrid proposal is denoted by VNS-GRASP instead of GRASP-VNS.

For the VNS versions, we use three neighbourhoods: swap-intra, swap-inter, and rebuild. The swap-intra move consists of the interchange of two POIs of a single solution itinerary. The swap-inter move consists of the interchange of two POIs of two different itineraries of the solution. These two moves are applied with the expectation of reducing the duration of some itinerary that makes room for the insertion of new POIs. The rebuild move consists of eliminating some POIs from a solution itinerary and inserting as many new POIs as possible. The POIs to be eliminated can be chosen at random, but the new POIs are inserted by the GRASP constructive procedure. This move is an interchange move extended by eliminating more than one POIs, and, even if only one POI is eliminated, the GRASP construction procedure can insert several new POIs. When a POI is to be inserted at a position in any itinerary, the new duration of the itinerary is computed. As mentioned above, the new POI is always inserted in the position with a minimum increase in the duration of the itinerary. The insertion of a new POI that means that the duration of the route would be greater than the limit is discarded. The number of POIs removed from an itinerary in the rebuild move is given by a parameter k_{max}.

In addition to their combinatorial complexity, another key factor shared by these contexts is their dynamic nature. In tourism, there may be last minute cancellations, weather changes, or changes in visitor preferences. Planning systems are therefore required not only to generate viable routes, but also to be able to adapt them quickly to changes in real-time [65]. The GRASP metaheuristic has a clear adaptive nature and can be used to complete a partial itinerary when it has already been realized, but the subsequent part of the itinerary is found unfeasible. On the other hand, *Adaptive VNS* (AVNS) offers a robust framework for optimization by changing the role to use the neighbourhoods. This approach combines the exploratory capabilities of the VNS method with learning mechanisms that dynamically adjust the search strategy based on observed performance [57]. Thus, it is possible to readjust already planned routes without having to start from scratch, maintaining both computational efficiency and solution quality.

5   Experimentation

In order to have a preliminary evaluation of the performance of the proposed AVNS-based approach for dynamic route optimization in an itinerary recommendation system, some experiments were conducted using real-world data from the island of Tenerife. The AVNS proposal is compared with a VNS with the same neighbourhoods and with GRASP. The size L of the RCL is set to 5 for all experiments. The dataset used consists of 149 POIs belonging to five main categories representative of tourist attractions: *Villages*, *Viewpoints*, *Trails*, *Beaches*, and *Restaurants*. In the category *Villages*, we include villages and towns of tourist interest. In the category *Beaches*, we also include other sea bath zones such as natural pools. In the category *Restaurants*, we also include other gastronomic resources. These POIs and also the viewpoints and trails are resources catalogued by the tourism regulator authority of the island; the Cabildo. The starting and end point for the itineraries is one of the many hotels in the capital of the island, the city of Santa Cruz. Each POI and the hotel have their true geographic coordinates and the travel times were computed using the corresponding tool of Open Street Map with OSMnx.

We consider TTDP model for stays of 5 days (itineraries) with a few additional constraints of those of the classical TOP model: limit of the diary number POIs for each category, maximum duration of the itineraries. The duration of each itinerary is limited to 6 h (540 min). Table 1 shows, for each category, the number of POIs in the category, the time duration interval (in minutes) of the category, and the maximum number of POIs of the category in each itinerary. The duration of the visit to every POI is randomly generated in the interval shown in this table using a triangular probability distribution. The POIs score for the instance was also randomly generated with a truncated normal distribution with mean 3.4, standard deviation 0.9, truncation interval [0,5] and rounded to 1 decimal.

Table 1. Data for the instance

Category	Number of POIs	Visit times	Max visits
Village	13	[60, 120]	2
Viewpoints	28	[10, 20]	5
Trails	20	[120, 180]	1
Beaches	51	[120, 240]	2
Restaurants	37	[60, 90]	1

We consider a dynamic TTDP where some disruption during the development of the itineraries implies their modification. We consider two types of disruption: 1) *Exclusion disruption*, when some of the POIs selected for the itinerary become unavailable at the planned instant, and 2) *Inclusion disruption*, when the tourist decides to visit also some POIs that are not included in the planned itinerary. In our preliminary experiments, we only tested single disruptions because there is one unavailable POI, or because only one new POI is required to visit. The moment of disruption can be on any of the days of the stay, at the beginning of the route, or during the itinerary (for instance, just at the moment of arriving at the unavailable POI or before). When a disruption is detected, a new recommendation has to be offered to the user, to the tourist. The new instance or the new problem, without the unavailable POIS definitively excluded for the selections and the mandatory POIs already included in the initial itineraries, can be solved by the same procedures. However, when a disruption occurs, we propose to use a new initial solution directly obtained from the previous recommendation. If the disruption is due to the unavailability of a POI we replace the POI by waiting time. If the disruption is due to the requirement to visit a new POI, this POI is inserted at the best position of the itinerary, and the following POIs of that itinerary are eliminated until the duration requirement is verified. The initial recommendations for the 5 itineraries in all the experiments included between 30 and 40 of the 150 POIs, so there were wide margins for changes.

We compare the performance of three algorithmic procedures: 1) apply the AVNS from the modified recommendation, 2) apply the nonadaptive VNS from the modified recommendation, and 3) apply the VNS-GRASP hybrid approach from an empty proposal. The performance indicators of the new proposal for the corresponding scenarios were the total score of the recommendation, the total duration of the itineraries and the computational time of the algorithms. Note that in addition to obtaining low computational times for the entire algorithm execution, sometimes tourists need an immediate answer on how to continue the current day itinerary at the moment of disruption.

In order to test the performances of the proposals we apply the VNS-GRASP hybrid for the instance obtained with the above parameters and use the series of 5 itineraries provided by the algorithm. Tables 2 and 3 show the results obtained for 10 cases where For the unavailability disruptions, we select at random a POI of one of the itinerary as unavailable POI. Then we apply the three possible strategies to get a new recommendation: the proposed AVNS, the VNS and the VNS-GRASP hybrid. The results in Table 2 show the results in which we consider rebuild moves eliminating only one POI from the solution ($k_{max} = 1$), where in Table 3 the results are in which the rebuild moves remove up to 3 POIs from the itineraries ($k_{max} = 1$).

Table 2. Comparative results. A single unavailable POI. $k_{max} = 1$

POI	AVNS			VNS			VNS-GRASP		
	Score	Duration	Time	Score	Duration	Time	Score	Duration	Time
2	160.5	2607.24	10.44	152.7	2592.13	8.91	150.6	2574.47	15.74
13	160.9	2633.22	9.06	154.4	2618.02	10.79	147.8	2582.07	12.27
2	157.9	2591.39	9.56	151.7	2612.97	6.79	149.1	2591.53	15.81
13	160.2	2613.95	9.05	155.4	2622.95	8.91	152.5	2550.92	13.48
13	160.8	2637.07	9.61	154.0	2619.02	10.79	151.0	2507.77	15.17
80	160.2	2622.37	10.41	153.4	2586.73	12.34	146.9	2602.14	16.02
80	157.7	2629.34	12.53	152.6	2572.60	8.62	147.3	2492.40	15.39
2	157.5	2629.38	11.62	151.8	2621.03	9.02	152.2	2561.54	15.55
13	161.0	2631.17	11.14	156.7	2546.01	15.70	154.1	2664.64	15.26
80	156.5	2628.59	6.36	153.8	2621.67	14.38	153.3	2661.40	11.26
Aver.	159.32	2622.37	9.98	153.65	2601.31	10.63	150.48	2578.89	14.59

Table 3. Comparative results. A single unavailable POI. $k_{max} = 3$

POI	AVNS			VNS			VNS-GRASP		
	Score	Duration	Time	Score	Duration	Time	Score	Duration	Time
70	160.5	2639.02	26.58	151.7	2615.29	12.93	148.6	2611.33	36.81
70	160.4	2609.69	15.87	150.7	2586.04	15.05	154.7	2625.96	37.96
70	160.6	2629.35	25.42	152.3	2650.00	19.01	150.8	2584.35	33.09
55	160.2	2640.30	26.48	153.0	2594.34	22.20	150.5	2625.44	26.92
70	159.7	2610.35	20.66	151.9	2605.05	15.57	151.7	2603.94	38.10
55	159.7	2611.64	17.67	152.2	2609.08	16.83	151.0	2557.56	37.82
70	159.5	2634.24	21.11	154.9	2670.00	25.93	149.6	2667.41	36.31
80	160.8	2631.01	24.99	156.3	2665.05	31.17	155.9	2639.67	37.53
11	160.3	2621.55	20.15	153.0	2578.28	12.70	152.9	2569.78	36.80
55	158.1	2588.86	18.37	151.9	2634.81	15.06	150.0	2597.80	35.43
Aver.	159.98	2621.60	21.73	152.79	2620.79	18.64	151.57	2608.32	35.68

Similarly, to test the performances of the three proposals with mandatory POI disruption, we also consider the same instance and solution with 5 itineraries that above. Tables 4 and 5 show the results obtained for 10 cases of a single POI requirement by the tourist. We select at random a POI not in the recommended itineraries and consider it as a mandatory POI at the beginning of the corresponding day. Then we also apply the three possible strategies to get a new recommendation: the proposed AVNS, the VNS and the VNS-GRASP hybrid. Analogously, the results in Table 4 show the results for $k_{max} = 1$ and in Table 5 are the results $k_{max} = 1$.

Table 4. Comparative results. A single required POI. $k_{max} = 1$

POI	AVNS			VNS			VNS-GRASP		
	Score	Duration	Time	Score	Duration	Time	Score	Duration	Time
124	153.8	2641.20	10.20	149.8	2599.68	15.61	147.9	2558.48	9.65
100	158.0	2599.27	8.95	151.6	2621.94	12.03	147.7	2544.66	15.31
140	160.1	2615.91	12.39	152.0	2562.98	4.71	148.4	2490.20	16.96
144	160.6	2636.85	10.06	153.5	2601.03	11.19	153.8	2578.07	10.97
140	158.7	2605.33	11.06	153.4	2624.50	8.36	152.8	2606.98	15.34
33	159.4	2620.71	8.91	152.0	2597.15	10.63	149.6	2536.29	15.03
40	157.5	2607.12	10.10	152.7	2636.17	7.74	151.3	2550.08	15.38
25	158.5	2629.72	9.62	153.2	2572.72	9.53	147.5	2530.41	10.47
12	157.0	2623.95	10.14	154.0	2605.33	11.59	148.1	2492.79	15.60
28	157.7	2617.89	8.99	152.5	2608.17	6.50	147.1	2627.68	14.65
Aver.	158.13	2619.80	10.04	152.47	2602.97	9.79	149.42	2551.56	13.93

Table 5. Comparative results. A single required POI. $k_{max} = 3$

POI	AVNS			VNS			VNS-GRASP		
	Score	Duration	Time	Score	Duration	Time	Score	Duration	Time
49	157.1	2627.75	17.65	152.4	2616.16	34.87	149.0	2607.92	38.56
18	157.3	2661.65	16.59	155.0	2647.57	37.44	154.0	2656.14	37.70
40	158.0	2631.99	20.71	154.1	2621.39	22.47	153.5	2555.63	38.10
93	150.9	2593.80	21.91	152.0	2612.58	19.41	154.9	2650.02	38.54
113	158.1	2644.88	23.45	153.2	2640.87	37.92	152.4	2587.57	37.06
51	148.1	2617.50	21.70	154.2	2566.12	35.24	150.1	2534.04	38.73
9	156.1	2630.12	20.23	152.5	2612.90	18.77	151.3	2546.82	28.96
116	158.0	2639.89	20.19	155.0	2614.73	33.28	150.8	2549.53	38.37
81	159.7	2639.11	16.03	153.2	2595.23	22.01	147.8	2635.96	35.91
98	155.6	2588.45	19.80	148.9	2579.01	13.32	150.2	2604.45	28.31
Aver.	155.89	2627.51	19.83	153.05	2610.66	27.47	151.4	2592.81	36.02

6 Conclusions

We propose a dynamic approach to dealing with dynamic tourist itinerary recommendations. This approach is based on an Adaptive VNS that dynamically
modify the probability of selecting each neighbourhood. The dynamism in the
recommendation arises from the possibility that, during the itinerary development, a disruption may alter the circumstances in two ways. In this preliminary
analysis, we consider two basic but realistic disruptions. First, a POI included
in the recommended itineraries becomes unavailable. Second, the tourist decides
to visit a POI not included in the recommended itineraries. We propose, instead
of solving the problem anew considering the new circumstances, to apply a
VNS algorithm from the previous recommended itineraries modified to take into
account the new requirements. We consider a standard VNS and an adaptive version hybridized with GRASP whose constructive procedure is used in a rebuild
move inside the VNS.

The main preliminary conclusions of the experiments are as follows. Both
VNS versions, the standard VNS and the adaptive VNS from the previous recommendation as starting solution provide better solutions in less computational
time than the GRASP-VNS. Second, the adaptive VNS has better performance
than the non adaptive approach.

Future works will be orientated towards the application of other data science
tools in order to adjust the values of the parameters stated at any case and
their dynamic adaptation to the circumstances. A very simple extension of this
research will include the unavailability of a set of POIs (for instance, a category
of POIs like beaches or trials due to weather conditions) or the requirements
to visit more than one POI not initially included in the recommendation; even
a disruption with several unavailable POIs and several mandatory POIs at the

same time. Other realistic changes in the circumstances could be tested. For example, the enlargement or shortening of the duration of some visit or some translation between POIs.

References

1. Aarts, E., Korst, J.: Simulated Annealing and Boltzmann Machines: A Stochastic Approach to Combinatorial Optimization and Neural Computing. Wiley (1989)
2. Archetti, C., Hertz, A., Speranza, M.G.: Metaheuristics for the team orienteering problem. J. Heuristics **13**, 49–76 (2007). https://doi.org/10.1007/s10732-006-9004-0
3. de Armas, J., Moreno-Pérez, J.: A survey on variable neighborhood search for sustainable logistics. Algorithms **18**(1), 38 (2025). https://doi.org/10.3390/a18010038
4. Beneich, C., Douiri, S.M.: Dynamic multi-compartment vehicle routing problem: formulation and algorithm. In: Farhaoui, Y., Hussain, A., Saba, T., Taherdoost, H., Verma, A. (eds.) Artificial Intelligence, Data Science and Applications, pp. 100–105. Springer, Cham(2024). https://doi.org/10.1007/978-3-031-48573-2_15
5. Brimberg, J., Salhi, S., Todosijević, R., Urošević, D.: Variable neighborhood search: The power of change and simplicity. Comput. Oper. Res. **155**, 106221 (2023). https://doi.org/10.1016/J.COR.2023.106221
6. Cebeci, E., Yücel, E., Koç, C.: The home health care routing with heterogeneous electric vehicles and synchronization. OR Spectrum (2024). https://doi.org/10.1007/s00291-024-00765-z
7. Chao, I.M., Golden, B.L., Wasil, E.A.: The team orienteering problem. Eur. J. Oper. Res. **88**(3), 464–474 (1996). https://doi.org/10.1016/0377-2217(94)00289-4
8. Chen, B., Qu, R., Bai, R., Laesanklang, W.: A variable neighborhood search algorithm with reinforcement learning for a real-life periodic vehicle routing problem with time windows and open routes. RAIRO - Oper. Res. **54**(5), 1467–1494 (2020). https://doi.org/10.1051/ro/2019080
9. Chen, S., Yin, Y., Sang, H., Deng, W.: A hybrid grasp and VND heuristic for vehicle routing problem with dynamic requests. Egyptian Inf. J. **29**, 100638 (2025)
10. Delahaye, D., Chaimatanan, S., Mongeau, M.: Simulated annealing: from basics to applications. In: Gendreau, M., Potvin, J.-Y. (eds.) Handbook of Metaheuristics. ISORMS, vol. 272, pp. 1–35. Springer, Cham (2019). https://doi.org/10.1007/978-3-319-91086-4_1
11. Dontas, M., Sideris, G., Manousakis, E.G., Zachariadis, E.E.: An adaptive memory matheuristic for the set orienteering problem. Eur. J. Oper. Res. **309**(3), 1010–1023 (2023). https://doi.org/10.1016/j.ejor.2023.02.008
12. Erdem, M.: Optimisation of sustainable urban recycling waste collection and routing with heterogeneous electric vehicles. Sustain. Cities Soc. **80**, 103785 (2022). https://doi.org/10.1016/j.scs.2022.103785
13. Feo, T.A., Resende, M.G.C.: Greedy randomized adaptive search procedures. J. Global Optim. **6**(2), 109–133 (1995). https://doi.org/10.1007/BF01096763
14. García-Martínez, C., Rodriguez, F.J., Lozano, M.: Genetic algorithms. In: Martí, R., Pardalos, P.M., Resende, M.G. (eds.) Handbook of Heuristics, pp. 627–662. Springer, Cham (2025).https://doi.org/10.1007/978-3-032-00385-0_28

15. Gavalas, D., Konstantopoulos, C., Mastakas, K., Pantziou, G.: A survey on algorithmic approaches for solving tourist trip design problems. J. Heuristics **20**(3), 291–328 (2014). https://doi.org/10.1007/s10732-014-9242-5
16. Gavalas, D., Konstantopoulos, C., Mastakas, K., Pantziou, G., Vathis, N.: Efficient heuristics for the time dependent team orienteering problem with time windows. In: Gupta, P., Zaroliagis, C. (eds.) ICAA 2014. LNCS, vol. 8321, pp. 152–163. Springer, Cham (2014). https://doi.org/10.1007/978-3-319-04126-1_13
17. Ghobadi, F., Divsalar, A., Jandaghi, H., Nozari, R.B.: An integrated recommender system for multi-day tourist itinerary. Appl. Soft Comput. **149**, 110942 (2023). https://doi.org/10.1016/j.asoc.2023.110942
18. Goldberg, D.E.: Genetic Algorithm in Search, Optimization and Machine Learning, Addison. Wesley Publishing Company (1989)
19. González-Navasa, C., Moreno Pérez, J.A., Brito, J.: A hybrid metaheuristic for a tourist route recommender. In: Quesada-Arencibia, A., Affenzeller, M., Moreno-Díaz, R. (eds.) EUROCAST 2024. vol. 15172, pp. 221–235. Springer, Cham (2025). https://doi.org/10.1007/978-3-031-82949-9_21
20. González-Navasa, C., Moreno Pérez, J.A., Brito-Santana, J., Alonso-Afonso, H.: Recommendation of tourist itineraries with dependence on transport time. Transp. Res. Procedia **71**, 77–84 (2023). https://doi.org/10.1016/j.trpro.2023.11.060
21. Gunawan, A., Lau, H.C., Vansteenwegen, P.: Orienteering problem: a survey of recent variants, solution approaches and applications. Eur. J. Oper. Res. **255**(2), 315–332 (2016). https://doi.org/10.1016/j.ejor.2016.04.059
22. Gunawan, A., Lau, H.C., Vansteenwegen, P., Lu, K.: Well-tuned algorithms for the team orienteering problem with time windows. J. Oper. Res. Soc. **68**(8), 861–876 (2017). https://doi.org/10.1057/s41274-017-0244-1
23. Halder, S., Lim, K.H., Chan, J., Zhang, X.: A survey on personalized itinerary recommendation: From optimisation to deep learning. Appl. Soft Comput. **152**, 111200 (2024). https://doi.org/10.1016/j.asoc.2023.111200
24. Hamid, R.A., et al.: How smart is e-tourism? A systematic review of smart tourism recommendation system applying data management. Comput. Sci. Rev. **39**, 100337 (2021). https://doi.org/10.1016/j.cosrev.2020.100337
25. Hansen, P., Mladenović, N., Brimberg, J., Pérez, J.A.M.: Variable neighborhood search. In: Gendreau, M., Potvin, J.-Y. (eds.) Handbook of Metaheuristics. ISORMS, vol. 272, pp. 57–97. Springer, Cham (2019). https://doi.org/10.1007/978-3-319-91086-4_3
26. Hansen, P., Mladenović, N.: Variable neighborhood search: principles and applications. Eur. J. Oper. Res. **130**, 449–467 (2001). https://doi.org/10.1016/S0377-2217(00)00100-4
27. Hansen, P., Mladenović, N., Moreno Pérez, J.A.: Variable neighbourhood search: methods and applications. 4OR **6**, 319–360 (2008). https://doi.org/10.1007/s10288-008-0089-1
28. Hansen, P., Mladenović, N., Moreno Pérez, J.A.: Variable neighbourhood search: methods and applications. Ann. Oper. Res. **175**, 367–407 (2010). https://doi.org/10.1007/s10479-009-0657-6
29. Jewpanya, P., Nuangpirom, P., Pitjamit, S., Nakkiew, W.: Optimized travel itineraries: combining mandatory visits and personalized activities. Algorithms **18**(2) (2025). https://doi.org/10.3390/a18020110
30. Kalatzantonakis, P., Sifaleras, A., Samaras, N.: A reinforcement learning-variable neighborhood search method for the capacitated vehicle routing problem. Expert Syst. Appl. **213**, 118812 (2023). https://doi.org/10.1016/j.eswa.2022.118812

31. Karimi-Mamaghan, M., Mohammadi, M., Meyer, P., Karimi-Mamaghan, A.M., Talbi, E.G.: Machine learning at the service of meta-heuristics for solving combinatorial optimization problems: a state-of-the-art. Eur. J. Oper. Res. **296**(2), 393–422 (2022). https://doi.org/10.1016/j.ejor.2021.04.032
32. Khamsing, N., Chindaprasert, K., Pitakaso, R., Sirirak, W., Theeraviriya, C.: Modified ALNS algorithm for a processing application of family tourist route planning: a case study of Buriram in Thailand. Computation **9**(2), 23 (2021). https://doi.org/10.3390/computation9020023
33. Khodadadian, M., Divsalar, A., Verbeeck, C., Gunawan, A., Vansteenwegen, P.: Time dependent orienteering problem with time windows and service time dependent profits. Comput. Oper. Res. **143**, 105794 (2022). https://doi.org/10.1016/j.cor.2022.105794
34. Liao, Z., Zheng, W.: Using a heuristic algorithm to design a personalized day tour route in a time-dependent stochastic environment. Tour. Manage. **68**, 284–300 (2018). https://doi.org/10.1016/j.tourman.2018.03.012
35. Lin, S.W., Yu, V.F.: A simulated annealing heuristic for the multiconstraint team orienteering problem with multiple time windows. Appl. Soft Comput. **37**, 632–642 (2015). https://doi.org/10.1016/j.asoc.2015.08.058
36. Lin, S.W., Yu, V.F.: Solving the team orienteering problem with time windows and mandatory visits by multi-start simulated annealing. Comput. Ind. Eng. **114**, 195–205 (2017). https://doi.org/10.1016/j.cie.2017.10.020
37. Lipowski, A., Lipowska, D.: Roulette-wheel selection via stochastic acceptance. Phys. A Stat. Mech. Appl. **391**(6), 2193–2196 (2012). https://doi.org/10.1016/j.physa.2011.12.004
38. Liu, W., Dridi, M., Ren, J., El Hassani, A.H., Li, S.: A double-adaptive general variable neighborhood search for an unmanned electric vehicle routing and scheduling problem in green manufacturing systems. Eng. Appl. Artif. Intell. **126**, 107113 (2023). https://doi.org/10.1016/j.engappai.2023.107113
39. Luo, X.G., Liu, X.R., Ji, P.L., Shang, X.Z., Zhang, Z.L.: Trip planning for visitors in a service system with capacity constraints. Comput. Oper. Res. **148**, 105974 (2022). https://doi.org/10.1016/j.cor.2022.105974
40. Migliorini, S., Quintarelli, E., Gambini, M., Belussi, A., Carra, D.: Sequence recommendations for groups: a dynamic approach to balance preferences. Inf. Syst. **108**, 102023 (2022). https://doi.org/10.1016/j.is.2022.102023
41. Mladenović, N., Hansen, P.: Variable neighborhood search. Comput. Oper. Res. **24**, 1097–1100 (1997). https://doi.org/10.1016/S0305-0548(97)00031-2
42. Mladenović, N.: A variable neighborhood algorithm-a new metaheuristic for combinatorial optimization. In: Papers presented at Optimization Days, vol. 12 (1995)
43. Polacek, M., Benkner, S., Doerner, K.F., Hartl, R.F.: A cooperative and adaptive variable neighborhood search for the multi depot vehicle routing problem with time windows. Bus. Res. **1**(2), 207–218 (2008). https://doi.org/10.1007/bf03343534
44. Resende, M.G.C., Ribeiro, C.C.: Greedy randomized adaptive search procedures: advances and extensions. In: Gendreau, M., Potvin, J.-Y. (eds.) Handbook of Metaheuristics. ISORMS, vol. 272, pp. 169–220. Springer, Cham (2019). https://doi.org/10.1007/978-3-319-91086-4_6
45. Ricci, F.: Recommender systems in tourism. In: Xiang, Z., Fuchs, M., Gretzel, U., Höpken, W. (eds.) Handbook of e-Tourism, pp. 457–474. Springer, Cham (2022) https://doi.org/10.1007/978-3-030-48652-5_26
46. Ronconi, D.P., Manguino, J.L.: Grasp and VNS approaches for a vehicle routing problem with step cost functions. Ann. Oper. Res. **350**(1), 37–62 (2025). https://doi.org/10.1016/j.ejor.2016.04.059

47. Ruiz-Meza, J., Brito, J., Montoya-Torres, J.R.: A GRASP-VND algorithm to solve the multi-objective fuzzy and sustainable Tourist Trip Design Problem for groups. Appl. Soft Comput. **131**, 109–716 (2022). https://doi.org/10.1016/j.asoc.2022.109716

48. Ruiz-Meza, J., Montoya-Torres, J.R.: Tourist trip design with heterogeneous preferences, transport mode selection and environmental considerations. Ann. Oper. Res. **305**(1), 227–249 (2021). https://doi.org/10.1007/s10479-021-04209-7

49. Ruiz-Meza, J., Montoya-Torres, J.R.: A systematic literature review for the tourist trip design problem: extensions, solution techniques and future research lines. Oper. Res. Perspect. **9**, 100–228 (2022). https://doi.org/10.1016/j.orp.2022.100228

50. Sancho Núñez, J.C., Gómez-Pulido, J.A., Ramírez, R.: Machine learning applied to tourism: a systematic review. Wiley Interdisc. Rev. Data Min. Knowl. Discov. **14** (2024).https://doi.org/10.1002/widm.1549

51. Queiroz dos Santos, J.P., de Melo, J.D., Duarte Neto, A.D., Aloise, D.: Reactive search strategies using reinforcement learning, local search algorithms and variable neighborhood search. Expert Syst. Appl. **41**(10), 4939–4949 (2014). https://doi.org/10.1016/j.eswa.2014.01.040

52. Santos-Peñate, D.R., Moreno-Pérez, J., Rodríguez, C.C., Suárez-Vega, R.: A mathematical model and GRASP for a tourist trip design problem. In: In: Moreno-Díaz, R., Pichler, F., Quesada-Arencibia, A. (eds.) EUROCAST 2022. LNCS, vol. 13789, pp. 112–120. Springer, Cham (2022). https://doi.org/10.1007/978-3-031-25312-6_13

53. Sarkar, J.L., Majumder, A., Panigrahi, C.R., Roy, S., Pati, B.: Tourism recommendation system: a survey and future research directions. Multimedia Tools Appl. **82**(6), 8983–9027 (2023). https://doi.org/10.1007/s11042-022-12167-w

54. Souffriau, W., Vansteenwegen, P., Berghe, G.V., Oudheusden, D.V.: The multiconstraint team orienteering problem with multiple time windows. Transp. Sci. **47**(1), 53–63 (2013). https://doi.org/10.1287/trsc.1110.0377

55. Stützle, T., Ruiz, R.: Iterated local search. In: Martí, R., Pardalos, P.M., Resende, M.G. (eds.) Handbook of Heuristics, pp. 779–807. Springer, Cham (2025). https://doi.org/10.1007/978-3-032-00385-0_8

56. Sylejmani, K., Abdurrahmani, V., Ahmeti, A., Gashi, E.: Solving the tourist trip planning problem with attraction patterns using meta-heuristic techniques. Inf. Technol. Tourism **26**(4), 633–678 (2024). https://doi.org/10.1007/s40558-024-00297-w

57. Sze, J.F., Salhi, S., Wassan, N.: A hybridisation of adaptive variable neighbourhood search and large neighbourhood search: application to the vehicle routing problem. Expert Syst. Appl. **65**, 383–397 (2016). https://doi.org/10.1016/j.eswa.2016.08.060

58. Sze, J.F., Salhi, S., Wassan, N.: An adaptive variable neighbourhood search approach for the dynamic vehicle routing problem. Comput. Oper. Res. **164**, 106531 (2024). https://doi.org/10.1016/j.cor.2024.106531

59. Talbi, E.G.: Machine learning into metaheuristics: a survey and taxonomy. ACM Comput. Surv. (CSUR) **54**(6), 129:1-129:32 (2021). https://doi.org/10.1145/3459664

60. Theeraviriya, C., Ruamboon, K., Praseeratasang, N.: Solving the multi-level location routing problem considering the environmental impact using a hybrid metaheuristic. Int. J. Eng. Bus. Manag. **13** (2021). https://doi.org/10.1177/18479790211017353

61. Vansteenwegen, P., Souffriau, W., Berghe, G.V., Oudheusden, D.V.: Metaheuristics for tourist trip planning. In: Sörensen, K., Sevaux, M., Habenicht, W., Geiger,

M.J. (eds.) Metaheuristics in the Service Industry. Lecture Notes in Economics and Mathematical Systems, pp. 15–31. Springer, Heidelberg (2009). https://doi.org/10.1007/978-3-642-00939-6_2

62. Vansteenwegen, P., Souffriau, W., Van Oudheusden, D.: A detailed analysis of two metaheuristics for the team orienteering problem. In: Stützle, T., Birattari, M., Hoos, H.H. (eds.) SLS 2009. LNCS, vol. 5752, pp. 110–114. Springer, Heidelberg (2009). https://doi.org/10.1007/978-3-642-03751-1_9

63. Vansteenwegen, P., Souffriau, W., Oudheusden, D.V.: The orienteering problem: a survey. Eur. J. Oper. Res. **209**(1), 1–10 (2011). https://doi.org/10.1016/j.ejor.2010.03.045

64. Verbeeck, C., Vansteenwegen, P., Aghezzaf, E.H.: Solving the stochastic time-dependent orienteering problem with time windows. Eur. J. Oper. Res. **255**(3), 699–718 (2016). https://doi.org/10.1016/j.ejor.2016.05.031

65. Wang, T., Gu, Y., Wang, H., Wu, G.: Adaptive variable neighborhood search algorithm with metropolis rule and tabu list for satellite range scheduling problem. Comput. Oper. Res. **170**, 106757 (2024). https://doi.org/10.1016/j.cor.2024.106757

66. Wu, L., Gu, T., Chen, Z., Zeng, P., Liao, Z.: Personalized day tour design for urban tourists with consideration to CO2 emissions. Chinese J. Popul. Resour. Environ. **20**(3), 237–244 (2022). https://doi.org/10.1016/j.cjpre.2022.09.004

67. Yahyaoui, H., Kaabachi, I., Krichen, S., Dekdouk, A.: Two metaheuristic approaches for solving the multi-compartment vehicle routing problem. Oper. Res. Int. J. **20**(4), 2085–2108 (2018). https://doi.org/10.1007/s12351-018-0403-4

68. Yalcin, G.D., Malta, H., Saylik, S.: A new mathematical model and a heuristic algorithm for the tourist trip design problem under new constraints: a real-world application. OPSEARCH **60**(4), 1703–1730 (2023). https://doi.org/10.1007/s12597-023-00678-5

69. Zeinab Aliahmadi, S., Jabbarzadeh, A., Hof, L.A.: A multi-objective optimization approach for sustainable and personalized trip planning: A self-adaptive evolutionary algorithm with case study. Expert Syst. Appl. **261**, 125412 (2025). https://doi.org/10.1016/j.eswa.2024.125412

General Variable Neighborhood Search for the Multi-source Capacitated Facility Location Problem with Customer Incompatibilities and Maximum Budget Service Constraints

Joan Serrano Roig[1]($\boxtimes$) , Anna Martínez-Gavara[1] , and Miguel Reula[2]

[1] Departamento de Estadística e Investigación Operativa, Universidad de Valencia, C/ Doctor Moliner, 50, Burjassot 46100, Valencia, Spain
joan.serrano@uv.es, gavara@uv.es

[2] Matemáticas para la Economía y la Empresa, Universidad de Valencia, Avinguda dels Tarongers, S/N, Algirós 46022, València, Spain
miguel.reula@uv.es

Abstract. We address a variant of the Capacitated Facility Location Problem in which each customer can be assigned to multiple facilities, subject to customer incompatibility constraint where, specifically, no two incompatible customers may be served by the same facility. In this paper, we extend the model by introducing a maximum budget service constraint, requiring that each customer must be assigned only to facilities within a predefined maximum unitary cost. This variant has practical applications where proximity and compatibility are critical, such as emergency logistics, last-miler delivery, and distribution of social services. Given the $\mathcal{NP}$-hard nature of the problem, we propose a metaheuristic based on General Variable Neighborhood Search (GVNS), which integrates adaptive neighborhood structures and efficient local search procedures. Extensive computational experiments confirm the robustness and efficiency of the proposed method, particularly in scenarios with tight budget service restrictions.

Keywords: Facility Location · Multi-Source · Customer Incompatibilities · Metaheuristic · Maximum Budget

1 Introduction

Facility Location Problems (FLPs) are classical optimization problems concerned with determining the optimal placement of a subset of facilities to efficiently serve a set of customers [1,2]. The quality of a facility location configuration depends on the specific objective function, which typically aims to minimize

---

A. Martínez-Gavara and M. R. Marin—These authors contributed equally to this work.

S. Cavero et al. (Eds.): ICVNS 2025, LNCS 16256, pp. 46–61, 2026.
https://doi.org/10.1007/978-3-032-19582-1_4

the overall service cost. FLPs generally involve two types of costs: a fixed cost incurred when opening a facility (setup cost), and a variable cost associated with assigning customers to open facilities (connection or transportation cost). For instance, in warehouse location planning, the objective is to minimize the sum of the fixed costs of establishing warehouses and the transportation costs of delivering goods to customers. Given the high costs associated with establishing and operating infrastructure networks, careful preliminary planning is essential to reduce long-term maintenance and relocation expenses.

Two fundamental variants have been widely studied in the literature, the uncapacitated Facility Location Problem (UFLP), in which facilities can serve an unlimited number of customers, and the Capacitated Facility Location Problem (CFLP), where each facility has a predefined capacity limiting the number of customers it can serve. UFLP and CFLP are $\mathcal{NP}$-hard optimization problems (Cornuéjols et al. [3]) and have been extensively investigated due to their practical relevance in applications such as sensor network optimization (Furuta et al. [4]), data clustering (Meira et al. [5]), placement of routers and caches (Guha et al. [6]), agglomeration of traffic or data (Andrews and Zhang [7]), locate switching centers in a communications network and police stations in a highway system (Hakimi [8]), emergency humanitarian logistics [9–11], hospital emergency (Sitepu et al. [12]), or wireless transmission infrastructure (Whitaker and Hurley [13]).

A recent theoretical review by Pelegrín [14] discusses several variants, including the Simple Facility Location Problem (SFLP), which restricts each customer to be served by a single supplier. Their work further extends the classical model to incorporate constraints such as customer incompatibilities and co-location requirements, modeled respectively as "cannot link" and "must link" conditions. In the context of the CFLP, Maia et al. [15] introduced the Multi-Source Capacitated Facility Location Problem with Customer Incompatibilities (MS-CFLP-CI), a variant in which multiple facilities may serve each customer, but specific pairs of customers cannot be assigned to the same facility due to incompatibility constraints. These incompatibilities reflect real-world scenarios such as rival customers who prohibit shared suppliers or customers requiring mutually exclusive product types. To solve this complex problem, Maia et al. [15] developed a portfolio of metaheuristic techniques, including a MineReduce approach based on data mining for problem size reduction, a Greedy Randomized Adaptive Search Procedure (GRASP), a permutation-coded evolutionary algorithm, and a multi-start greedy algorithm. Their experiments, conducted on a newly proposed dataset of 30 benchmark instances (with up to 3000 facilities), showed that MineReduce achieved the best results across most scenarios.

In 2024, Ceschia and Schaerf [16] proposed an advanced local search method guided by a multi-neighborhood Simulated Annealing (SA) framework to address the MS-CFLP-CI. Their approach introduces novel neighborhood structures, including facility closure and reopening mechanisms, and improves search space reduction by focusing on solutions with at most two suppliers per customer. The algorithm is initialized using a greedy method and fine-tuned via F-Race for parameter optimization. Their results demonstrated significant improvements over previously proposed methods by Maia et al. [15]. Furthermore, their contribution includes a complementary dataset, an exact mathematical formulation to solve small instances, and a complete open-source implementation of their solution method.

In many real-world applications, ensuring proximity between suppliers and customers is crucial for maintaining service quality and operational efficiency. For instance, in the pharmaceutical distribution sector, pharmacies often operate under exclusive contracts that prevent them from sharing suppliers with direct competitors. In such cases, customer incompatibilities are a natural choice for modeling. However, another critical factor is response time, particularly in urgent or high-turnover environments where delayed deliveries can result in medication shortages. To mitigate this risk, it becomes essential to restrict assignments to nearby facilities, thereby guaranteeing short delivery times. Incorporating a maximum budget service constraint into the facility location model addresses this need directly. It ensures that customers are only served by facilities within a predefined geographical or operational radius, achieving a balance between exclusivity and logistical feasibility. This additional constraint improves the applicability of the model in contexts where minimizing delivery times is essential.

To the best of our knowledge, no previous studies have incorporated a maximum budget service constraint into the MS-CFLP-CI. While many extensions of the CFLP exist, addressing aspects such as capacity limits, customer incompatibilities, and multi-sourcing, none explicitly limit the cost between customers and facilities in the presence of these other constraints. Motivated by this gap, we introduce a new variant: the MS-CFLP-CI with Maximum Budget Service, which integrates all three components: capacity, customer incompatibility, and service restrictions, into a single, unified model. Note that in the literature review, we find the maximum budget service constraint as a maximum distance between facilities and customers. For example, Li and Lu [17] studies a Facility Location Problem with Maximum Distance (FLP-MD), proving that it is $\mathcal{NP}$-hard and equivalent to the set cover problem. Another related work is due to Villegas et al. [18], where the authors address a biobjective uncapacitated FLP involving cost and coverage.

To solve the proposed MS-CFLP-CI with Maximum Budget Service constraint, we develop a metaheuristic based on General Variable Neighborhood Search (GVNS). This framework combines a deterministic Variable Neighborhood Descent (VND) for local intensification with a shaking phase designed to promote diversification and escape from local optima. GVNS has proven to be a robust and adaptable metaheuristic across a wide range of combinatorial optimization problems. It has been successfully applied to facility location problems, such as the Prize-Collecting Capacity-Constrained Connected Facility Location Problem [19] and the Hard Capacitated k-Facility Location Problem [20]. Therefore, we propose a GVNS method to efficiently explore the complex solution space of our problem variant, incorporating neighborhood structures and feasibility-preserving moves that respect capacity, incompatibility, and service constraints.

The main contributions of this paper are as follows: (i) the extension of the MS-CFLP-CI by introducing a maximum budget service constraint, and the proposal of a corresponding mathematical programming model; (ii) the implementation of the mathematical model using Gurobi to solve small-sized instances; (iii) the creation of a new data set of instances of varying sizes; and (iv) the development and implementation of the GVNS based metaheuristic to solve the MS-CFLP-CI with the maximum budget service constraint.

The rest of the paper is organized as follows. In Sect. 2, we formally define the problem and present its formulation as an integer programming method. In Sect. 3, we describe the algorithm based on the GVNS methodology. Section 4 presents the computational results, including a comparison with the solutions obtained from the exact mathematical model implemented using Gurobi. Finally, Sect. 5 concludes the paper and outlines directions for future research.

2 Mathematical Model

As previously defined, the MS-CFLP-CI consists of determining the set of facilities to be located to satisfy customer demand while satisfying incompatibility constraints. To incorporate a maximum budget service constraint, we formulate a Mixed Integer Linear Program (MILP) based on the MS-CFLP-CI model presented in Maia et al. [15] and Ceschia and Schaerf [16], providing a rigorous framework for analyzing, solving, and validating feasible solutions.

Let S and W be the number of customers and suppliers, respectively. We define the set of customers as $\mathcal{I} = \{1, \ldots, S\}$, where each $i \in \mathcal{I}$ has a demand d_i. The set of potential facilities is $\mathcal{J} = \{1, \ldots, W\}$, each with capacity s_j. A solution is feasible if it satisfies all customer demands, the facility capacity limits, and the incompatibility constraints, where any pair $(i_1, i_2) \in \Gamma$ cannot be assigned to the same facility, being Γ the set of pairs of incompatible customers. Before we dive into the model, we need to introduce the two types of costs used for the problem: c_{ij} and f_j represent the supply cost and the fixed cost, respectively. In addition, the supplier assignment and opening matrices are represented by x_{ij} and y_j. Based on the classical CFLP, the problem formulation for the multi-source (MS) variant is:

$$(\text{MS-CFLP}): \min \quad \sum_{i \in \mathcal{I}} \sum_{j \in \mathcal{J}} x_{ij} c_{ij} + \sum_{j \in \mathcal{J}} y_j f_j \tag{1}$$

$$\text{s.t.} \quad \sum_{j \in \mathcal{J}} x_{ij} = d_i \qquad \forall i \in \mathcal{I} \tag{2}$$

$$\sum_{i \in \mathcal{I}} x_{ij} \leq s_j y_j \qquad \forall j \in \mathcal{J} \tag{3}$$

$$0 \leq x_{ij} \leq d_i \qquad \forall i \in \mathcal{I}, \forall j \in \mathcal{J} \tag{4}$$

$$y_j \in \{0, 1\} \qquad \forall j \in \mathcal{J} \tag{5}$$

The objective function (1) minimizes the total cost of supplying customers and the fixed cost of opening facilities. Equations (2) and (3) ensure in this order that every customer is completely served and suppliers do not exceed their capacity to provide. We may note that constraint (3) also guarantees that no customer is assigned to a closed facility. In addition, equations (4) and (5) are the variables' domain.

The customer incompatibilities will be satisfied with the inclusion of the following constraints:

$$x_{i_1 j} = 0 \vee x_{i_2 j} = 0 \quad \forall (i_1, i_2) \in \Gamma, \forall j \in \mathcal{J}. \tag{6}$$

Equations (1)-(6) define the MS-CFLP-CI, which is not linear because of the incompatibility constraint (6), so we need to reformulate it. Some authors, such as Maia et al. [15], use the big M technique; instead, we define z_{ij} so that

$$z_{i_1 j} + z_{i_2 j} \leq y_j, \quad \forall (i_1, i_2) \in \Gamma, \tag{7}$$

$$0 \leq x_{ij} \leq d_i z_{ij} \quad \forall i \in \mathcal{I}, \forall j \in \mathcal{J}. \tag{8}$$

The reader may notice that equation (7) is the linearization of (6), and constraint (4) needs to be updated to (8) due to the definition of $z-$variables. A similar formulation is used in Ceschia and Schaerf [16], but in this model, x_{ij} defines the amount of goods instead of the percentage of the customers' demand.

Finally, as we consider that customers cannot be assigned to facilities beyond a certain service cost, $C_{\max}$, we include the last constraint (9),

$$z_{ij} = 0, \quad \forall i \in \mathcal{I}, \forall j \in \mathcal{J} : c_{ij} > C_{\max}. \tag{9}$$

Then, after some adaptation of the original model of the MS-CFLP, Equations (1)–(3) and (5)–(9) result in the MS-CFLP-CI with Maximum Budget Service constraint, named as `MS-CFLP-CI-MB`.

3   General Variable Neighborhood Search

For this problem, we propose a General Variable Neighborhood Search (GVNS) to solve the MS-CFLP-CI-MB. The pseudocode of the algorithm is presented in Algorithm 1. This proposal has four input parameters: k_{min}, k_{step}, k_{max}, and t_{max}. The first three mentioned control the value of the shaking number k, while t_{max} defines the maximum CPU time. The values of k_{min} and k_{max} are defined using the IRACE framework (see López-Ibáñez et al. [21]) and determine the percentage of facilities closed in the shake stage, from k_{min} to k_{max}, increasing it in k_{step} in each iteration. As k_{min} and k_{max} are variable, k_{step} must adapt to every different combination. To do it, an auxiliary variable n_{steps} is defined such that:

$$k_{step} = \frac{k_{max} - k_{min}}{n_{steps}}. \tag{10}$$

The pseudocode presented in Algorithm 1 starts by constructing an initial solution x (Step 2). While it is common in the VNS literature to generate this solution randomly, previous studies [22] have shown that more elaborate construction methods can improve the quality of the final solution. In our approach, the initial solution is generated using a constructive procedure described in Sect. 3.2, and subsequently improved (Step 3) using a Variable Neighborhood Descent (VND) procedure, as detailed in Sect. 3.4.

The main loop of the GVNS is repeated until the time limit $t_{\max}$ is reached (Steps 4–12). Initially, the neighborhood index k is set to $k_{\min}$. Starting from the first predefined neighborhood structure, the algorithm iterates until the maximum neighborhood $k_{\max}$ is reached. In each iteration, three main procedures are executed: Shake, VND, and Neighborhood Change.

Given a current solution x, the Shake procedure generates a new solution x' by perturbing x using the method defined in Sect. 3.3 (Step 7). The resulting solution x' is then improved using the VND procedure to obtain x'' (Step 8). The VND procedure is explained in Sect. 3.4. Finally, the Neighborhood Change method determines the next neighborhood to explore (Step 9). If x'' improves upon x in terms of the objective function value, the solution is updated ($x \leftarrow x''$), and the search restarts from the first neighborhood ($k \leftarrow k_{min}$). Otherwise, the algorithm moves to the next neighborhood ($k \leftarrow k + k_{step}$).

Algorithm 1. GVNS

1: **function** GVNS(k_{min},k_{step}, k_{max}, t_{max})
2:      $x \leftarrow$ CONSTRUCT_SOL
3:      $x \leftarrow$ VND(x)
4:      **repeat**
5:          $k \leftarrow k_{min}$
6:          **repeat**
7:              $x' \leftarrow$ SHAKE(x, k)
8:              $x'' \leftarrow$ VND(x')
9:              $x, k \leftarrow$ NEIGHCHANGE(x, x'', k)
10:          **until** $k = k_{max}$
11:          $t \leftarrow$ CpuTime()
12:      **until** $t > t_{max}$
13:      **return** x
14: **end function**

3.1   Solution Representation

We begin by describing the solution representation used in our heuristic, which enables efficient transformations between neighboring solutions during the search. Given the large-scale nature of the benchmark instances, ranging from 50 to 3000 suppliers, it was crucial to design a data structure that minimizes memory overhead and facilitates fast updates during each iteration.

Following the approach proposed in Ceschia and Schaerf [16], we represent the assignment matrix using two separate matrices where each row corresponds to a customer. The first column of both matrices records the number of suppliers assigned to each customer. The remaining columns in the first matrix identify the assigned facilities, while the second matrix stores the corresponding quantities delivered by each facility. This design avoids the need for exhaustive row-wise iteration, allowing direct access to relevant data during neighborhood evaluations.

Additionally, since some of the neighborhoods used in our GVNS algorithm are facility-orientated, we include a third matrix with one row for each facility, to store the number of customers it serves and their corresponding indices. This allows for efficient identification and modification of customer–supplier relationships from both customer and facility-based perspectives.

This representation not only ensures feasibility but also reduces the computational cost of neighborhood exploration, which is particularly important in large instances where quick feasibility checks are essential.

3.2   Construction Method

The construction phase is responsible for generating a feasible initial solution that satisfies the constraints of the MS-CFLP-CI-MB problem. In a standard VNS implementation, the initial solution is constructed at random; however, recent studies [22,23] show that an initial solution with significantly improved quality can speed the convergence of the algorithm. Our construction method follows a greedy randomized approach consisting of three main stages: (i) facility opening, (ii) customer assignment, and (iii) solution completion.

The first step, based on **opening facilities**, involves selecting an initial set of suppliers to open. The percentage of facilities to open, denoted by the parameter θ, is predefined and is fine-tuned in the numerical experiments. To guide the selection, we compute a vector that reflects, for each supplier j, the number of customers i for which supplier j is the best candidate according to a cost metric $F(\cdot,\cdot)$. Then, the greedy function $g_1(j)$ is defined as the number of customers for whom j is the best supplier:

$$g_1(j) = \left| \left\{ i \in \mathcal{I} : j = \arg \min_{l \in \mathcal{J} : c_{il} \leq C_{\max}} F(i,l) \right\} \right|, \tag{11}$$

where $F(i,l)$ is one of the following cost metrics: pure transportation cost, $F_1(i,l)$, adjusted cost with fixed cost, $F_2(i,l)$, and normalized cost, $F_3(i,l)$ such that

$$F_1(i,l) = c_{il}, \quad F_2(i,l) = c_{il} + f_l, \quad F_3(i,l) = c_{il} + \frac{f_l}{s_l} \tag{12}$$

Then, depending on the chosen cost metric, the algorithm computes $g_1(j)$ for all $j \in \mathcal{J}$, and uses this score to define a discrete probability distribution that favors suppliers with higher $g_1(j)$. A set of θ suppliers is then sampled from this distribution and marked as open. At this point, no customer assignments are made.

Once a set of suppliers is open, the next step is to **assign customers** to them. Due to the maximum service constraint, some customers have limited assignment options, which may lead to infeasibility if they are not carefully selected. To prioritize customers with fewer available suppliers, we define a selection probability inversely proportional to the number of feasible assignments. Customers with fewer available facilities have a higher chance of being selected. The algorithm iteratively selects a customer from this distribution, assigns them to the closest available and compatible supplier, and fills the supplier's capacity. Assignments are made only if the supplier has sufficient residual capacity and the customers are mutually compatible. Additionally, this constructive phase operates under a non-splitting rule: the demand of each customer must be fully allocated to a single supplier.

Finally, suppose the current set of open facilities is not enough to cover all the demands of the customers. In that case, the remaining unassigned customers are assigned using a **completion phase**. In this phase, a Restricted Candidate List (RCL) is constructed containing the feasible pairs (i, j) where i is an unassigned customer and j is its closest supplier available. For each feasible pair (i, j), a greedy function $g_2(i, j) = F(i, j)$ is computed, where $F(i, j)$ is defined above, and can be one of the cost metrics functions F_1, F_2, or F_3. The RCL is constructed by selecting only those feasible pairs (i, j) whose greedy value does not exceed a threshold τ, defined as:

$$\tau = F_{min} + \alpha \left(F_{max} - F_{min} \right), \tag{13}$$

where

$$F_{min} = \min_{(i,j)} F(i, j),$$
$$F_{max} = \max_{(i,j)} F(i, j) \tag{14}$$

for all feasible pairs (i, j). The parameter $\alpha \in [0, 1]$, configured in Sect. 4, controls the greediness of the selection. If $\alpha = 0$, the most promising pair is selected, obtaining a purely greedy selection. When $\alpha = 1$, all feasible pairs are included in the RCL, obtaining a random selection procedure. After implementing IRACE, its optimal value is $\alpha = 0.5561$. At each iteration, one pair $(i, j) \in RCL$ is randomly selected and added to the solution. If supplier j is not yet open, it is activated, and customer i is assigned. Before continuing the process, once a supplier j becomes available, it is filled using the customer assignment phase described above. This process is repeated until all customer demands are satisfied, ensuring the feasibility of the final constructed solution.

3.3   Shake Procedure

The shaking procedure is a key component of the **GVNS** algorithm, designed to diversify the search and escape from local optima. It perturbs the current solution to generate a structurally different neighbor, which is then improved by using the VND procedure. Our implementation follows a destruction–repair strategy, where the extent of the perturbation is controlled by the parameter k, which ranges from k_{min} to $k_{\max}$.

In the destruction phase, a percentage k of the currently open facilities is randomly closed. The value of k increases from k_{min} to k_{max} in increments of k_{step} units. These parameters were tuned through numerical experiments, resulting in the configuration $k_{min} = 0.2935$, $k_{max} = 0.9993$, and $k_{step} = 0.3529$ (obtaining $n_{steps} = 2$, see the definition of k_{step} in Equation (10)). Closing a facility forces the reassignment of its customers, introducing variability into the solution and the exploration of new promising regions in the search space. Once the destruction is complete, the repair phase is applied to restore feasibility and rebuild the solution. Customers affected by the closed facility are reassigned to new suppliers using our constructive procedure described in Sect. 3.2. This reconstruction process combines efficiency with solution quality. As a result, the repair phase not only restores feasibility but also guides the solution toward promising areas of the solution space. In Duarte et al. [24] and SánchezOro et al. [25], the authors explored more sophisticated shaking strategies, such as intensified shaking, designed to achieve a better balance between diversification and intensification.

3.4   Variable Neighborhood Descent

Variable Neighborhood Descent (VND) is a variant of the VNS metaheuristic that deterministically explores several neighborhood structures considering the first improvement local search strategy. Its effectiveness comes from the simple fact that different neighborhoods rarely share the same local minimum. By systematically changing neighborhoods, VND can effectively escape the trap of local optima. The algorithm starts with the first neighborhood $(\mathcal{N}_1)$ and, at each iteration, proceeds to the next neighborhood if no improving solution is found. If an improving move is performed, the search restarts from $\mathcal{N}_1$. VND stops when the last neighborhood $\mathcal{N}_{\ell_{max}}$ is explored with no improvement. We propose a total of five neighborhoods $(\ell_{max} = 5)$, which have been categorized depending on their main target: customer-orientated or facility-orientated. The order of the neighborhoods is not arbitrary but follows a simple strategy: from the fastest to the slowest. To determine this order, we computed a performance ratio defined as the number of iterations per second for each neighborhood, using a random sample of instances with varying sizes and characteristics. The sample consisted of three benchmark instances. For each instance, we performed eight experiments with different random parameter configurations, resulting in a total of 24 experiments. In each run, we recorded the number of moves executed (whether the solution improved or not) and the total computational time for each neighborhood. The aggregated performance across all experiments determined the final neighborhood order: Close Facility $(\mathcal{N}_1)$, Close-Open Facility $(\mathcal{N}_2)$, Split Customer $(\mathcal{N}_3)$, Group Customer $(\mathcal{N}_4)$, and Swap Incompatible Customers $(\mathcal{N}_5)$. For the sake of simplicity, the detailed comparison results are not included.

3.4.1 Facility-Orientated Neighborhood Structures

In this section, we define the two neighborhoods that are focused on generating moves by closing and opening suppliers. Given a solution x, let $\mathcal{J}_o$ and $\mathcal{J}_c$ denote the disjoint set of open and closed suppliers, respectively, such that the complete set of facilities is

$$\mathcal{J}(x) = \mathcal{J}_o(x) \cup \mathcal{J}_c(x). \tag{15}$$

The *Close Facility* move, denoted as $Close(u)$, consists of closing an open supplier u and reassigning its customers to other already open facilities, provided that feasibility is maintained. The neighborhood $\mathcal{N}_1(x)$ associated with this move is defined as:

$$\mathcal{N}_1(x) = \{x' : x' = x \oplus Close(u), \quad u \in \mathcal{J}_o\}. \tag{16}$$

To guide this move, we consider to select the supplier with the following three criteria: (i) the fewest assigned customers, (ii) the highest cost, or (iii) the highest cost per assigned customer. The parameter $\mu = \{1, 2, 3\}$ determines the criteria used: (i), (ii) or (iii), respectively. We evaluate the different strategies via IRACE, obtaining that the best configuration is $\mu = 2$.

The *Close-Open Facility* move, denoted as $CloseOpen(u, v)$, reassigns the load from an open supplier $u \in \mathcal{J}_o(x)$ to a closed one $v \in \mathcal{J}_c(x)$. The neighborhood associated is formally defined as follows:

$$\mathcal{N}_2(x) = \{x' : x' = x \oplus CloseOpen(u, v), \quad u \in \mathcal{J}_o, v \in \mathcal{J}_c\}. \tag{17}$$

Since customers assigned to the same facility are, by definition, compatible, this move can be efficiently evaluated by identifying a closed supplier with sufficient residual capacity. To select the facility u to be closed, we evaluate different strategies tuned via IRACE, including selecting the open supplier with the fewest customers or the one with the highest cost. The *Close Facility* and the *Close-Open Facility* moves share the parameter μ, and the optimal value is the same for both. See the experimental setup in Sect. 4. For selecting a new facility v, we construct a cost vector that ranks the closed supliers in ascending order based on their cumulative cost to the affected customers. The algorithm attempts to assign customers to the cheapest feasible supplier in this list. If the move results in an improvement in the objective function, it is accepted; otherwise, the next candidate in the vector is evaluated.

3.4.2 Customer-Orientated Neighborhood Structures

This section presents the set of neighborhood structures focused on customer-based modifications. A total of three customer-orientated neighborhoods have been defined, and their aim is to restructure the assignment of the demands through customer splitting, grouping, or swap incompatible customers.

The *Split Customer*, denoted as $Split(i)$, is designed to intensify the search by splitting the demand of a customer i among multiple suppliers. Following the empirical finding of Ceschia and Schaerf [16], where the majority of customers were served by a single supplier and only a small fraction by multiple, we allow at most ρ suppliers active per customer. The parameter ρ is an integer in $[1, 4]$, determined by IRACE, and defined in Sect. 4, and obtaining $\rho = 4$ as the best configuration. The neighborhood $\mathcal{N}_3(x)$ contains all the solutions obtained after applying $Split(i)$ feasible moves. The algorithm selects the most expensive customer who has not yet reached the maximum allowed number of splits. Then, the demand of the customer is redistributed by reallocating quantities from less cost-effective suppliers to the highest ones.

The *Group Customer* move, denoted as $Group(i)$, aims to reassign the demand of customer i to a single, closer supplier, merging the split demand whenever possible. If the customer is already fully assigned to one facility, the algorithm seeks a cost-reducing reassignment to a closer open facility. To prioritize moves, a vector is constructed by evaluating the total assignment cost for each customer: $\sum_{j \in \mathcal{J}} c_{ij} x_{ij} \quad \forall i \in \mathcal{I}$, and the customers are sorted in descending order. The move is applied to the most expensive customer, if no improvement is found the next candidate is evaluated. After every successful move, the cost list is updated, and the process restarts. Therefore, this neighborhood is formally defined as

$$\mathcal{N}_4(x) = \{x' : x' = x \oplus Group(i), \quad i \in \mathcal{I}\}. \tag{18}$$

Incompatibility constraints poses a challenge in the search space, as they significantly limit the number of feasible moves. To address this issue, we introduce the *Swap Incompatible Customers* move, denoted as $Swap(i_1, i_2)$, which directly operates on pairs of incompatible customers $(i_1, i_2) \in \Gamma$. The procedure starts by selecting the most expensive customer, followed by the random selection of one of its incompatible customers. If both can be feasibly reassigned (partially or fully) to each other's facilities without violating any constraints, the swap move is evaluated and performed only if it results in an improvement in the objective function. Formally, the associated neighborhood is defined as

$$\mathcal{N}_5(x) = \{x' : x' = x \oplus Swap(i_1, i_2), \quad (i_1, i_2) \in \Gamma\}. \tag{19}$$

4   Computational Experiments

This section is organized as follows. First, we describe the characteristics of the benchmark instances used and the strategy adopted to adapt them for our problem setting. Next, we detail the parameter tuning process. Finally, we present the computational results sobtained with the exact solver (Gurobi) and the proposed GVNS algorithm.

As mentioned in the Introduction, this problem is inspired by the MS-CFLP-CI in Maia et al. [15]. The original dataset, created for MESS 2020+1 and composed of 30 instances, is available on the website https://www.ants-lab.it/mess2020/#competition and was tested to guarantee feasible solutions despite the incompatibility constraints. The dataset created by Maia et al. [15] served as the starting point for generating our own instances, including the maximum budget service parameter, $C_{\max}$. The new dataset and the code developed can be found in https://github.com/JoanSerranoOR/ICVNS-2025.

One of the key aspects of algorithm design is the careful selection of parameters that guide the search process and improve solution quality. To achieve this, we employed the Iterated Racing for Automatic Algorithm Configuration (IRACE) framework developed by López-Ibáñez et al. [21].

For the parameter tuning, we used five benchmark instances, each evaluated under three different values of $C_{\max}$. Table 1 summarizes the parameters tuned with IRACE, together with their optimal values, ranges, and data types. The parameter α controls the greediness of the construction phase (see Sect. 3.2), θ defines the percentage of suppliers to be opened in the facility opening step of the construction phase (Sect. 3.2), and $F(\cdot, \cdot)$ specifies the cost metric function used to evaluate candidate facilities in the opening step, where $F(i, l)$ define the greedy function g_1. As described in Sect. 3.4.1, μ determines the criterion for selecting suppliers in the facility-orientated neighborhoods. The parameters k_{min}, k_{max}, and n_{steps} are defined in Sect. 3 and used to compute k_{step}; together, they control the number and intensity of each shake iteration in the GVNS. Finally, ρ is the maximum number of active suppliers per customer (see Sect. 3.4.2). As mentioned previously, we have used the original instances defined in the literature and added the C_{max} parameter to them. The final configuration of these instances is detailed in Table 2.

Table 1. Parameters' configuration.

Name	Data type	Initial range	Value	Name	Data type	Initial range	Value
α	Double	$[0.2, 0.6]$	0.5561	k_{min}	Float	$[0.1, 0.5]$	0.2935
θ	Integer	$[0.2, 0.6]$	0.4790	k_{max}	Float	$[0.6, 1.0]$	0.9993
$F(\cdot, \cdot)$	Integer	$[1, 3]$	3	n_{steps}	Integer	$[1, 20]$	2
μ	Integer	$[1, 3]$	2	ρ	Integer	$[1, 4]$	4

The MS-CFLP-CI-MB was implemented with Gurobi for the exact model and the GVNS algorithm coded in C++. Both implementations were executed on a system equipped with an AMD Ryzen 5 7530U processor and 16 GB of RAM. Tables 3 and 4 report the computational results obtained with the integer linear programming (ILP) model solved by Gurobi and with the proposed General VNS algorithm. Table 3 presents the comparison for the set of instances where Gurobi was able to reach the optimal solution, reporting z as the optimal objective function value. Table 4 shows the results for instances where Gurobi did not reach optimality, in this case, z represents the best solution found by the ILP model.

The remaining columns are as follows: $t[s]$, the computational time in seconds; *gap*, the deviation of each method with respect to the best solution found in the experiment; and *#Best*, which indicates if the corresponding method achieved the best result. Each instance was executed 10 times with the GVNS; in this case, *min* represents the best solution obtained and *mean* the average objective value across runs. If Gurobi runs out of time before reaching a feasible solution, the value z is marked as "–". The execution time limit for Gurobi is set to 3600 seconds, while the General VNS is allocated a time limit of $10 \cdot \sqrt{W}$ seconds, consistent with previous studies on the problem without the budget service constraint [15, 16]. Both time values are reported in the $t[s]$ column of each table (Tables 3 and 4).

Table 2. Configuration of each instance

Inst.	W	S	C_{max}	Inst.	W	S	C_{max}	Inst.	W	S	C_{max}
wlp01	50	115	24	wlp11	800	2020	14	wlp21	75	172	24
wlp02	100	253	18	wlp12	900	2159	15	wlp22	175	428	18
wlp03	150	345	19	wlp13	1000	2305	13	wlp23	275	694	16
wlp04	200	479	16	wlp14	1200	2927	12	wlp24	450	1128	15
wlp05	250	601	15	wlp15	1400	3445	14	wlp25	650	1619	14
wlp06	300	705	14	wlp16	1600	4067	15	wlp26	850	2007	16
wlp07	400	1012	14	wlp17	1800	4373	14	wlp27	1100	2847	12
wlp08	500	1277	14	wlp18	2000	4908	12	wlp28	1500	3474	14
wlp09	600	1483	13	wlp19	2500	5882	15	wlp29	1900	4522	15
wlp10	700	1733	15	wlp20	3000	7800	16	wlp30	2750	6965	13

The results in Tables 3 and 4 show that Gurobi was able to reach optimality in 9 out of 30 instances, all with 50 and 300 suppliers, but failed to prove optimality for medium and large instances within a time limit of 3600 s. In contrast, the proposed GVNS consistently provided high-quality solutions with an average gap below 12% and in substantially less time. This advantage was particularly evident in the largest instances, where Gurobi failed to find any feasible solution, whereas GVNS always delivered competitive results, demonstrating both robustness and scalability.

Table 3. Comparison between Gurobi and GVNS for instances solved to optimality.

Inst.	ILP				GVNS				
	z	t [s]	gap	#Best	min	mean	t [s]	gap	#Best
wlp01	**28879.0**	0.49	–	1	31036.0	31402.1	70.71	0.07	0
wlp02	**53023.0**	4.05	–	1	59973.0	62065.3	100.00	0.13	0
wlp03	**64329.0**	41.40	–	1	71553.0	71931.0	122.47	0.11	0
wlp04	**84633.0**	80.23	–	1	93694.0	94419.3	141.42	0.11	0
wlp05	**103829.0**	637.92	–	1	114513.0	115328.4	158.11	0.10	0
wlp06	**111499.0**	1679.63	–	1	123193.0	123938.6	173.21	0.10	0
wlp21	**38067.0**	1.28	–	1	41287.0	41679.6	86.60	0.08	0
wlp22	**74469.0**	149.59	–	1	81633.0	82209.5	132.29	0.10	0
wlp23	**119074.0**	2943.51	–	1	131946.0	132652.2	165.83	0.11	0
Summary	753113.3	615.34	–	9	832031.1	83958.4	127.85	0.10	0

Table 4. Comparison between Gurobi and GVNS for instances where no optimal value is found.

Inst.	ILP				GVNS				
	z	t [s]	gap	#Best	min	mean	t [s]	gap	#Best
wlp07	**162268**	3600		1	183966.0	186801.6	200.00	0.13	0
wlp08	**187882**	3600		1	207967.0	208590.5	223.61	0.11	0
wlp09	**218235**	3600		1	241004.0	241479.8	244.95	0.10	0
wlp10	**246380**	3600		1	269785.0	271054.4	264.58	0.09	0
wlp11	**288203**	3600		1	320234.0	321499.5	282.84	0.11	0
wlp12	**296473**	3600		1	327723.0	328717.0	300.00	0.11	0
wlp13	**314061**	3600		1	345409.0	346442.8	316.23	0.10	0
wlp14	**393812**	3600		1	435169.0	436802.0	346.41	0.11	0
wlp15	–	3600	–	0	**504731.0**	510419.7	374.17	0.00	1
wlp16	–	3600	–	0	**617148.0**	634123.5	400.00	0.00	1
wlp17	–	3600	–	0	**641345.0**	654601.4	424.26	0.00	1
wlp18	–	3600	–	0	**730524.0**	736084.4	447.21	0.00	1
wlp19	–	3600	–	0	**896965.0**	920336.3	500.00	0.00	1
wlp20	–	3600	–	0	**1101294.0**	1121517.6	547.72	0.00	1
wlp24	**168634**	3600		1	187615.0	188128.8	212.13	0.11	0
wlp25	**229619**	3600		1	254607.0	255645.5	254.95	0.11	0
wlp26	**289534**	3600		1	320723.0	321577.1	291.55	0.11	0
wlp27	**388335**	3600		1	428514.0	430873.0	331.66	0.10	0
wlp28	–	3600	–	0	**523257.0**	531915.1	387.30	0.00	1
wlp29	–	3600	–	0	**699554.0**	732144.1	435.89	0.00	1
wlp30	–	3600	–	0	**1015221.0**	1027857.6	524.40	0.00	1
Summary	–	3600	–	12	488226.4	495552.9	348.09	0.06	9

5   Conclusions and Future Work

In this work, we introduced the MS-CFLP-CI-MB, a novel extension of the multi-source capacitated facility location problem with customer incompatibilities by incorporating a maximum budget service constraint. This addition significantly reinforces the applicability of the model in scenarios where both proximity and exclusivity are decisive. We proposed a mixed-integer linear programming formulation to solve small-scale instances and developed a General Variable Neighborhood Search (GVNS) to address larger cases. Computational experiments demonstrated that Gurobi could solve the smallest instances to optimality, whereas the proposed GVNS consistently produced high-quality solutions with small optimality gaps and in substantially less time. The results highlight the robustness and scalability of GVNS, particularly in large and complex instances where exact methods either fail to reach optimality or cannot find feasible solutions within the time limit.

Future research will focus on improving the GVNS framework to solve the MS-CFLP-CI-MB model and integrating it into matheuristic approaches that combine exact methods with metaheuristic strategies.

References

1. Owen, S.H., Daskin, M.S.: Strategic facility location: a review. Eur. J. Oper. Res. **111**(3), 423–447 (1998). https://doi.org/10.1016/S0377-2217(98)00186-6
2. Klose, A., Drexl, A.: Facility location models for distribution system design. Eur. J. Oper. Res. **162**(1), 4–29 (2005). https://doi.org/10.1016/j.ejor.2003.10.031
3. Cornuéjols, G., Nemhauser, G., Wolsey, L.: The uncapicitated facility location problem. Technical report, Cornell University Operations Research and Industrial Engineering (1983)
4. Furuta, T., Sasaki, M., Ishizaki, F., Suzuki, A., Miyazawa, H.: A new clustering algorithm using facility location theory for wireless sensor networks. J. Oper. Res. Soc. of Jpn. **52**(4), 366–376 (2009)
5. Meira, L.A., Miyazawa, F.K., Pedrosa, L.L.: Clustering through continuous facility location problems. Theoret. Comput. Sci. **657**, 137–145 (2017). https://doi.org/10.1016/j.tcs.2016.10.001
6. Guha, S., Meyerson, A., Munagala, K.: Hierarchical placement and network design problems. In: Proceedings 41st Annual Symposium on Foundations of Computer Science, pp. 603–612 (2000). https://doi.org/10.1109/SFCS.2000.892328
7. Andrews, M., Zhang, L.: The access network design problem. In: Proceedings 39th Annual Symposium on Foundations of Computer Science (Cat. No.98CB36280), pp. 40–49 (1998). https://doi.org/10.1109/SFCS.1998.743427
8. Hakimi, S.L.: Optimum locations of switching centers and the absolute centers and medians of a graph. Oper. Res. **12**(3), 450–459 (1964). https://doi.org/10.1287/opre.12.3.450
9. Boonmee, C., Arimura, M., Asada, T.: Facility location optimization model for emergency humanitarian logistics. Int. J. Disaster Risk Reduct. **24**, 485–498 (2017). https://doi.org/10.1016/j.ijdrr.2017.01.017
10. Wang, W., Wu, S., Wang, S., Zhen, L., Qu, X.: Emergency facility location problems in logistics: status and perspectives. Transp. Res. Part E: Logistics Transp. Rev. **154**, 102465 (2021). https://doi.org/10.1016/j.tre.2021.102465
11. Caunhye, A.M., Nie, X., Pokharel, S.: Optimization models in emergency logistics: a literature review. Socioecon. Plann. Sci. **46**(1), 4–13 (2012). https://doi.org/10.1016/j.seps.2011.04.004
12. Sitepu, R., Puspita, F.M., Lestari, I., Yuliza, E., Octarina, S., et al.: Facility location problem of dynamic optimal location of hospital emergency department in Palembang. Sci. Technol. Indonesia **7**(2), 251–256 (2022). https://doi.org/10.26554/sti.2022.7.2.251-256
13. Whitaker, R.M., Hurley, S.: On the optimality of facility location for wireless transmission infrastructure. Comput. Ind. Eng. **46**(1), 171–191 (2004). https://doi.org/10.1016/j.cie.2004.01.002
14. Pelegrín, M.: New variants of the simple plant location problem and applications. Eur. J. Oper. Res. **306**(3), 1094–1108 (2023). https://doi.org/10.1016/j.ejor.2022.10.027
15. Maia, M.R., et al.: Metaheuristic techniques for the capacitated facility location problem with customer incompatibilities. Soft. Comput. **27**(8), 4685–4698 (2023). https://doi.org/10.1007/s00500-022-07600-z

16. Ceschia, S., Schaerf, A.: Multi-neighborhood simulated annealing for the capacitated facility location problem with customer incompatibilities. Comput. Ind. Eng. **188**, 109858 (2024). https://doi.org/10.1016/j.cie.2023.109858
17. Li, X., Lu, X.: The facility location problem with maximum distance constraint. Inf. Process. Lett. **184**, 106447 (2024). https://doi.org/10.1016/j.ipl.2023.106447
18. Villegas, J.G., Palacios, F., Medaglia, A.L.: Solution methods for the bi-objective (cost-coverage) unconstrained facility location problem with an illustrative example. Ann. Oper. Res. **147**(1), 109–141 (2006). https://doi.org/10.1007/s10479-006-0061-4
19. Leitner, M., Raidl, G.R.: Variable neighborhood search for a prize collecting capacity constrained connected facility location problem. In: 2008 International Symposium on Applications and the Internet, pp. 233–236 (2008). https://doi.org/10.1109/SAINT.2008.57
20. Mišković, S., Stanimirović, Z.: Variable neighborhood search based heuristics for the hard capacitated k-facility location problem. IPSI Bgd. Trans. Internet Res. **2017**, 1–8 (2017)
21. López-Ibáñez, M., Dubois-Lacoste, J., Cáceres, L.P., Birattari, M., Stützle, T.: The IRACE package: iterated racing for automatic algorithm configuration. Operat. Res. Perspect. **3**, 43–58 (2016). https://doi.org/10.1016/j.orp.2016.09.002
22. Duarte, A., Escudero, L.F., Martí, R., Mladenovic, N., Pantrigo, J.J., SánchezOro, J.: Variable neighborhood search for the vertex separation problem. Comput. Oper. Res. **39**(12), 3247–3255 (2012). https://doi.org/10.1016/j.cor.2012.04.017
23. Sánchez-Oro, J., Mladenović, N., Duarte, A.: General variable neighborhood search for computing graph separators. Optim. Lett. **11**(6), 1069–1089 (2014). https://doi.org/10.1007/s11590-014-0793-z
24. Duarte, A., Pantrigo, J.J., Pardo, E.G., Mladenovic, N.: Multi-objective variable neighborhood search: an application to combinatorial optimization problems. J. Global Optim. **63**(3), 515–536 (2014). https://doi.org/10.1007/s10898-014-0213-z
25. Sánchez-Oro, J., Pantrigo, J.J., Duarte, A.: Combining intensification and diversification strategies in VNS. An application to the vertex separation problem. Comput. Oper. Res. **52**, 209–219 (2014). https://doi.org/10.1016/j.cor.2013.11.008

Exact and Metaheuristic Methods for a No-Wait Two-Stage Hybrid Flow Shop with a Single Server

Fouad Medouar[(✉)] and Rachid Benmansour

National Institute of Statistics and Applied Economics (SI2M laboratory, INSEA),
Rabat, Morocco
{fmedouar,r.benmansour}@insea.ac.ma

Abstract. This paper studies the two-stage no-wait hybrid flow shop scheduling problem with a single server at the first stage, involving two parallel machines in stage one and a single machine in stage two. The presence of no-wait and server constraints makes the problem highly challenging. A mixed-integer programming (MIP) model is proposed to solve small instances optimally. For larger cases, we develop two constructive heuristics and metaheuristic approaches, including SA and several variants of GVNS. Computational experiments show that while the MIP is limited to small sizes, GVNS significantly outperforms the heuristics and SA, delivering high-quality solutions in short computation times. The best GVNS variant achieves the lowest average gaps relative to the lower bound, demonstrating the efficiency and robustness of the proposed methods.

Keywords: Scheduling · Flow shop · no-wait · Mixed integer programming · Variable neighborhood search

1 Introduction

Flow shop scheduling problem is a central topic in operations research, modeling production systems in which jobs must follow a predefined sequence across multiple machines. The main objective is often to optimize performance measures such as makespan, total completion time, or total tardiness.

This paper studies a two-stage hybrid flow shop problem in which the first stage consists of two identical parallel machines and the second stage contains a single machine. A single server is responsible for performing setup operations on the first stage, which implies that only one machine can undergo setup at a time. This results in a resource-sharing constraint, since the server is required for setups on both machines. As a consequence, the setup on a machine cannot be planned independently of the other, making their utilization strongly interdependent [15].

R. Benmansour—Contributing authors.

In addition, a no-wait constraint between stages is imposed; this means that once a job finishes processing on the first stage, it must immediately start on the second stage without any waiting time. Such restrictions arise in technologically sensitive industries, such as chemical production, steel manufacturing, or semiconductor processing, where intermediate storage is not possible and delays between stages may cause product deterioration or infeasibility [17]. These combined features substantially increase the complexity of the problem, which can be expressed in Graham's notation [5] as $F2(P_2, 1) \mid no\text{-}wait, S1 \mid C_{\max}$.

Scheduling problems with no-wait constraints have attracted significant research interest because of their theoretical and practical relevance [10]. While the classical no-wait flow shop has been extensively studied, the no-wait hybrid flow shop has received comparatively little attention. Comprehensive reviews of no-wait scheduling were provided by Allahverdi [1] and, more recently, by Utama et al. [16]. Nevertheless, several studies have investigated the two-stage no-wait hybrid flow shop problem. Chang [4] examined the case with parallel machines and separated setup/removal times, proposing heuristic algorithms and sequencing theorems. Wang and Liu [17] introduced a genetic algorithm for the same problem, where constructive heuristics were used for initialization and extensive experiments confirmed its efficiency. Moradinasab et al. [12] investigated a no-wait two-stage flexible flow shop with setup times using an adaptive imperialist competitive algorithm and a genetic algorithm, benchmarking their results against ant colony optimization. Zhong and Shi [20] studied a two-stage no-wait hybrid flow shop with inter-stage flexibility, proved the NP-hardness of a special case, and proposed two polynomial-time approximation algorithms.

Practical applications have also been reported. Liu et al. [9] applied a two-stage no-wait hybrid flow shop model to the scheduling of multi-airport departure flights, where a CPLEX-based approach proved effective on real data. Wang et al. [18] considered sequence-dependent setup times at the first stage, proposing a Hungarian-based lower bound, three heuristics, a branch-and-bound algorithm, and a tabu search. More recently, Xuan et al. [19] studied a hybrid flow shop with unrelated parallel machines and no-wait constraints, proposing a genetic simulated annealing algorithm that significantly outperformed existing heuristics.

To our knowledge, no solution method has yet been proposed for the two-stage hybrid flow shop problem with both no-wait and single-server constraints. To fill this gap, we make the following contributions:

- We develop a mixed-integer programming (MIP) formulation capable of solving small-sized instances to optimality within reasonable computational times.
- We design two constructive heuristics, alongside three variants of the well-known variable neighborhood search (VNS) algorithm, to efficiently address medium- and large-sized instances.
- We propose a theoretical lower bound to evaluate the quality of the solutions obtained by the proposed methods.

- We conduct a comparative analysis with a well-known metaheuristic from the literature, assessing both solution quality and computational efficiency.

The remainder of the paper is organized as follows. Section 2 provides a formal description of the problem together with the proposed mixed-integer programming formulation and a theoretical lower bound. Section 3 introduces two greedy constructive heuristics, and Sect. 4 presents describes the proposed general variable neighborhood search method along with the employed neighborhood structures. Section 5 then presents the simulated annealing algorithm. Section 6 reports the computational experiments, while Sect. 7 concludes the paper with final remarks.

2   Problem Statement and Mathematical Formulation

2.1   Problem Description

The problem $F2(P_2, 1) \,|\, no\text{-}wait, S1 \,|\, C_{\max}$ can be described as follows. There are two stages with two identical parallel machines at stage 1 and one machine at stage 2. At the first stage, a single server is responsible for performing the setup of each job on both machines. Since the server can only handle one setup at a time, only one machine can be set up at any given moment, in addition, we suppose that setup and processing operations on stage 1 are not separated. After completion on this stage, each job must be immediately processed on the second stage, which consists of a single machine. No intermediate waiting is allowed between the stages, reflecting the no-wait constraint. Each job requires a setup time on stage 1, and the processing times for stage 1 and stage 2 may differ across jobs.

The objective is to determine a feasible schedule that minimizes the makespan, i.e., the completion time of the last scheduled job on the second stage. The notation used throughout the paper is given in Table 1.

2.2   Mixed-Integer-Programming (MIP) formulation

In this section we propose a mixed-integer-programming (MIP) formulation for the problem $F2(P_2, 1) \,|\, no\text{-}wait, S1 \,|\, C_{\max}$. This model is based on completion time variables, which were introduced by Balas [2] for modeling job shop scheduling problem. We introduce dummy jobs processed on the server, represented by the set N_1, which represents the setup operations of jobs on the first stage. The actual processing time of jobs on stage 1 and stage 2 are represented by N_2 and N_3, respectively. Binary variables $x_{i,j}$ assign jobs to one of the parallel machines on stage 1, while $y_{j,k}$ and $z_{j,k}$ define the processing order of jobs on the first and second stages. The completion time of each job C_j is tracked for all $j \in N = N_1 \cup N_2 \cup N_3$, and the objective is to minimize the makespan $C_{\max}$.

Table 1. Parameters and decision variables

Notation	Description
Sets and parameters	
n	number of jobs
$N_1 : \{1, \ldots, n\}$	set of dummy jobs to be processed on the server
$N_2 : \{n+1, \ldots, 2n\}$	set of jobs to be processed on stage 1
$N_3 : \{2n+1, \ldots, 3n\}$	set of jobs to be processed on stage 2
$N : N_1 \cup N_2 \cup N_3$	
$M : \{1, 2\}$	set of machines on stage 1
s_j	setup of job j
p_{ij}	processing time of job j on stage i
p_j	equal to s_j if $j \in N_1$; p_{1j} if $j \in N_2$; p_{2j} if $j \in N_3$
B	a large positive constant
Decision variables	
$x_{i,j} \in \{0,1\}$	1 if job $j \in N_2$ is assigned to machine $i \in M$, 0 otherwise
$y_{j,k} \in \{0,1\}$	1 if job j precedes job k on stage 1, 0 otherwise, $j, k \in N_1$, $j \neq k$
$z_{j,k} \in \{0,1\}$	1 if job j precedes job k on stage 2, 0 otherwise, $j, k \in N_3$, $j \neq k$
C_j	Completion time of job j, $j \in N$
$C_{\max}$	Makespan

The proposed model is given as follows.

$$\min \; C_{\max} \tag{1}$$

$$C_{\max} \geq C_j \quad \forall j \in N_3 \tag{2}$$

$$\sum_{i \in M} x_{i,j} = 1 \quad \forall j \in N_2 \tag{3}$$

$$C_j \geq p_j + p_{j-n} + p_{j-2n} \quad \forall j \in N_3 \tag{4}$$

$$C_{k-n} \geq p_{k-n} + C_j - B\left(3 - x_{i,j} - x_{i,k} - y_{j-n,k-n}\right) \quad \forall i \in M, \; j, k \in N_2, \; j \neq k \tag{5}$$

$$y_{j,k} + y_{k,j} = 1 \quad \forall j, k \in N_1, \; j \neq k \tag{6}$$

$$C_k \geq p_k + C_j - B(1 - y_{j,k}) \quad \forall j, k \in N_1, \; j \neq k \tag{7}$$

$$C_k \geq p_k + C_j - B(1 - z_{j,k}) \quad \forall j, k \in N_3, \; j \neq k \tag{8}$$

$$z_{j,k} + z_{k,j} = 1 \quad \forall j, k \in N_3, \; j \neq k \tag{9}$$

$$C_j = C_{j-n} + p_j \quad \forall j \in N_3 \tag{10}$$

$$C_j = C_{j-2n} + p_{j-n} + p_j \quad \forall j \in N_3 \tag{11}$$

$$C_j \geq 0 \quad \forall j \in N \tag{12}$$

$$x_{i,j} \in \{0,1\} \quad \forall i \in M, j \in N_2 \tag{13}$$

$$y_{j,k} \in \{0,1\} \quad \forall j, k \in N_1 \tag{14}$$

$$z_{j,k} \in \{0,1\} \quad \forall j, k \in N_3 \tag{15}$$

The objective function (1) minimizes the makespan. Constraints (2) ensure that the makespan is not smaller than the completion time on stage 2 of all jobs. Constraints (3) guarantee that every job in stage 1 is assigned to exactly one machine. Inequalities (4) impose a lower bound on the completion time of each job in stage 2, accounting for its setup, stage 1, and stage 2 processing times. Constraints (5) enforce machine capacity limits on stage 1, if two jobs are assigned to the same machine, their processing order is determined by the binary variable $y_{j-n,k-n}$. Constraints (6) ensure that, for every pair of server jobs, exactly one precedes the other. Similarly, inequalities (7) maintain precedence feasibility on the server, while inequalities (8) and (9) impose the same logic on stage 2 jobs. Equalities (10) and (11) link completion times across stages by enforcing the no-wait condition, a job on stage 2 starts immediately after its predecessor jobs finish on the server and stage 1. Finally, constraints (12) enforce non-negativity of completion times, while (13)–(15) define the binary nature of the assignment and sequencing variables.

In the next section, we introduce valid lower bounds to serve as benchmarks for evaluating the MIP formulation and the heuristic and metaheuristic methods presented later.

2.3 Lower Bound

In this section, we present a lower bound for the problem $F2\,(P_2, 1)$ $|no\text{-}wait, S1|C_{\max}$. The construction of valid lower bound is an important step in the analysis of scheduling problems, since they provide theoretical benchmarks against which the quality of heuristic or exact solution methods can be measured. In particular, a good lower bound allows us to evaluate how far a computed schedule is from optimality and thus serves as a useful tool for assessing the performance of the proposed algorithms. The proposed lower bound is given as follows: $LB^* = \max(LB_1, LB_2, LB_3)$ where LB_1, LB_2 and LB_3 are defined in Propositions 1, 2 and 3 respectively.

Proposition 1.

$$LB_1 = \frac{\sum_{j \in N_1} (p_j + p_{j+n})}{2} + \min_{j \in N_3} p_j$$

Proof. On stage 1, each job occupies exactly one of the two machines for its setup and its processing; thus the total workload that must be executed on the two parallel machines is $\sum_{j \in N_1} p_j + \sum_{j \in N_2} p_j$. By a standard load (averaging) argument for two identical parallel machines, the earliest time at which all stage 1 work can be completed is at least $\frac{\sum_{j \in N_1} (p_j + p_{j+n})}{2}$. Under the no-wait constraint, the job that finishes last overall must still execute its stage 2 processing for some p_j with $j \in N_3$ after stage 1; hence

$$C_{\max} \geq \frac{\sum_{j \in N_1} (p_j + p_{j+n})}{2} + \min_{j \in N_3} p_j.$$

$\square$

Proposition 2.

$$LB_2 = \sum_{j \in N_1} p_j + \min_{j \in N_2} \left(p_j + p_{j+n} \right)$$

Proof. The total setup workload on stage 1 is $\sum_{j \in N_1} p_j$. Under the single-server constraint, all setups must be performed sequentially, so the stage 1 completion cannot be earlier than this sum. After the setups, consider any job $j \in N_2$. Its combined processing on stage 1 and stage 2 is $p_j + p_{j+n}$. Since the makespan must at least include one job's complete processing on both stages, we have

$$C_{\max} \geq \sum_{j \in N_1} p_j + \min_{j \in N_2} \left(p_j + p_{j+n} \right).$$

$\square$

Proposition 3.

$$LB_3 = \min_{j \in N_1} \left(p_j + p_{j+n} \right) + \sum_{j \in N_3} p_j$$

Proof. Consider the total workload on stage 2. Each job must execute its stage 2 processing after completing stage 1. Thus, the makespan is at least the minimal setup and processing of one job on stage 1 plus the sum of all stage 2 processing times:

$$C_{\max} \geq \min_{j \in N_1} \left(p_j + p_{j+n} \right) + \sum_{j \in N_3} p_j.$$

$\square$

3   Initialization Heuristics

To generate initial feasible solutions for the problem, we use two constructive heuristics. The first is based on the well-known NEH algorithm [13], which builds a job sequence by iteratively inserting each job at the position that minimizes the makespan. The second is a simpler greedy insertion heuristic, which constructs the sequence by appending jobs one by one according to their contribution to the makespan. The feasible solutions produced by these heuristics are then used as initial solutions to guide the metaheuristic algorithm described in Sect. 4.

3.1   NEH-Based Heuristic (H1)

The NEH-based heuristic constructs the initial sequence by first ranking jobs according to their total workload $TPT_j = s_j + p_{1j} + p_{2j}$ and then inserting them iteratively into a growing partial sequence. At each iteration, the considered job is tested in all possible positions, and the sequence that yields the minimum makespan is retained. The steps of the proposed heuristic are summarized in Fig. 1.

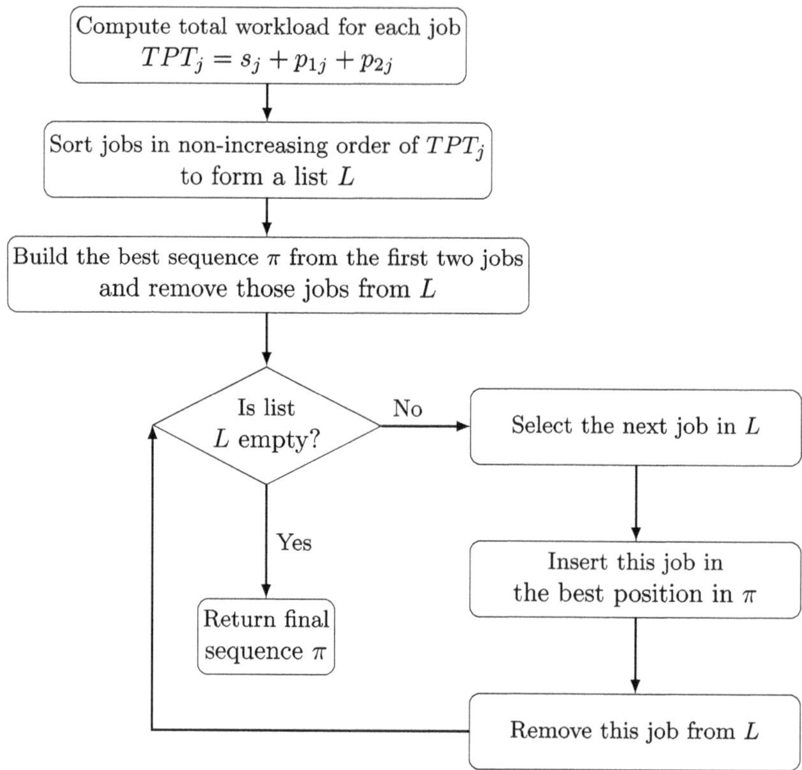

Fig. 1. Flowchart of the NEH-based heuristic H1.

3.2 Greedy Insertion Heuristic (H2)

The greedy insertion heuristic constructs the schedule incrementally by appending one job at a time to the end of the current partial sequence. At each step, a job is chosen as the one that minimizes the makespan when evaluated under the current sequence. The pseudocode of the proposed heuristic is given in Fig. 2.

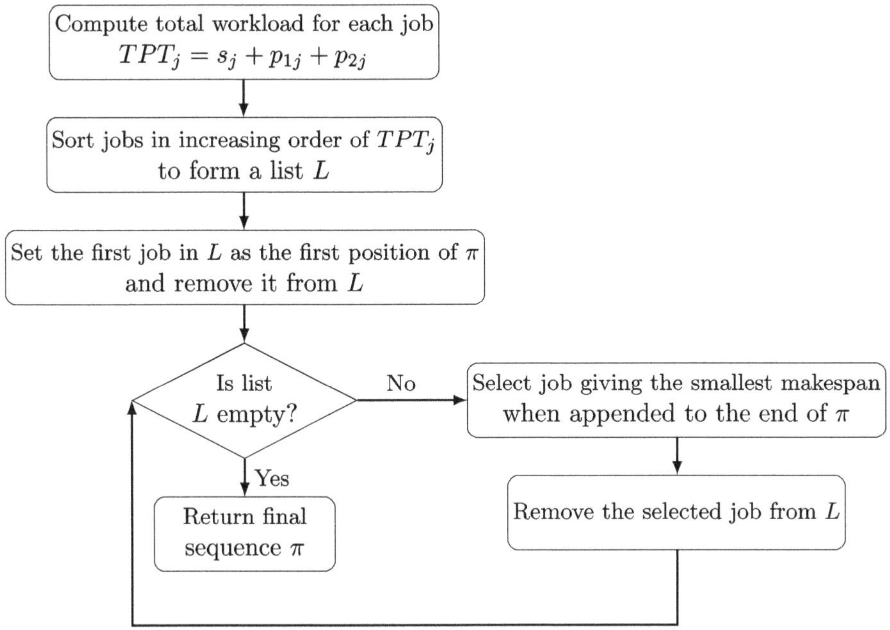

Fig. 2. Flowchart of the Greedy Insertion Heuristic H2.

4   General Variable Neighborhood Search

Variable Neighborhood Search (VNS), introduced by Mladenović and Hansen [11], is a widely used metaheuristic for solving complex combinatorial optimization problems. Its effectiveness has been demonstrated across various scheduling contexts [3,7]. The key idea of VNS is to systematically modify the neighborhood structure during the search in order to avoid premature convergence to local optima. Its basic form, referred to as Basic VNS (BVNS), alternates between a shaking step, which introduces diversification, and a local search step, which intensifies the search around the perturbed solution.

In this work, we develop a General Variable Neighborhood Search (GVNS) framework tailored to the problem $F2(P_2,1)\,|\,\text{no-wait},S1\,|\,C_{\max}$. The method integrates a shaking procedure to diversify the exploration of the search space and a Variable Neighborhood Descent (VND) scheme to intensify the search through systematically organized local improvements. Three variants of GVNS are considered, differing in the neighborhood-change strategy adopted within the VND procedure. Their general structure is illustrated in the flowchart of Fig. 3.

4.1   GVNS for the Problem $F2(P_2,1)\,|\,no\text{-}wait,S1\,|\,C_{\max}$

The overall algorithm begins with an initial solution obtained by selecting the best among two constructive heuristics. At each iteration, the current solution is perturbed through the shaking mechanism, after which a VND procedure is applied to intensify the search in its neighborhood. If an improved solution is found, it replaces the incumbent one and the shaking parameter is reset; otherwise, the neighborhood size used in the shaking phase is increased. This process continues until the time limit is reached. The complete structure of the GVNS metaheuristic is depicted in Fig. 3.

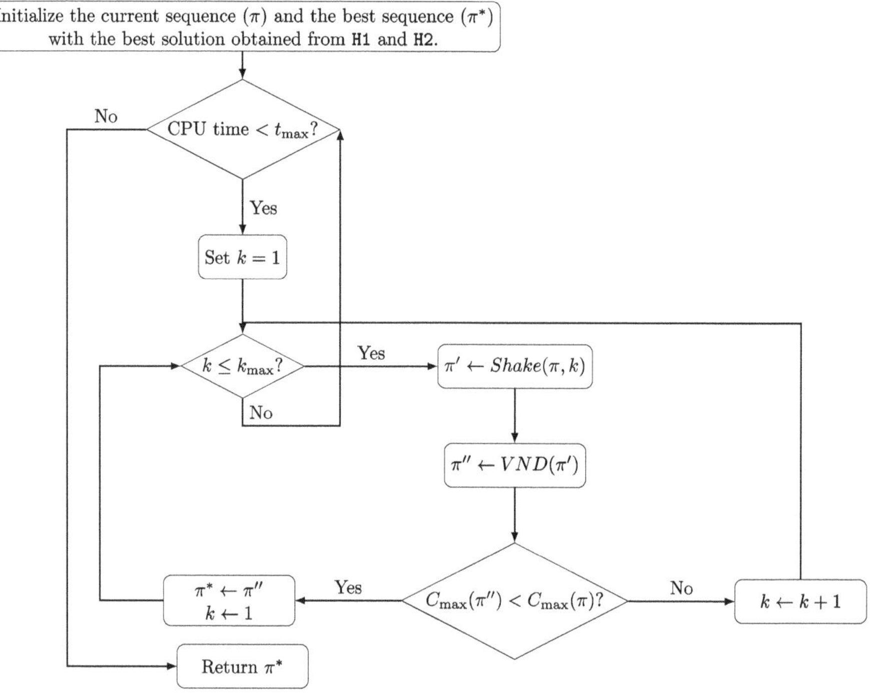

Fig. 3. Flowchart of the GVNS algorithm.

4.1.1 Neighborhood Structures

To explore the solution space, three neighborhood structures are employed: *swap*, which exchanges the positions of two jobs; *insert*, which removes a job from its position and reinserts it elsewhere; and *reverse*, which inverts the order of jobs within a selected subsequence. These neighborhoods are commonly used in scheduling due to their ability to induce both small and large perturbations, thereby enabling flexible navigation across different regions of the search space.

4.1.2 Shaking

The shaking procedure serves as the diversification mechanism of the GVNS algorithm. Starting from the current solution π, the procedure applies a sequence of random modifications based on the reverse neighborhood structure. At each perturbation, a subsequence of jobs is selected at random, and its order is reversed, yielding an intermediate solution that becomes the basis for the next perturbation. After performing k such perturbations, the resulting solution is returned as the shaken solution π'. This process allows the search to escape from local optima by exploring increasingly distant regions of the solution space as the shaking parameter grows.

4.1.3 Variable Neighborhood Descent (VND)

The intensification phase relies on a Variable Neighborhood Descent scheme, implemented in three variants: *sequential, cyclic,* and *pipe* VND. All versions operate on two neighborhood structures, namely *swap* and *insert,* but differ in the strategy used to navigate between them.

In the *sequential* strategy, the algorithm explores the neighborhoods in a fixed order. Starting from the shaken solution, it applies a local search in the first neighborhood; if an improvement is found, the search restarts from that neighborhood, otherwise it moves to the next one. The procedure stops when no improvement can be obtained from any neighborhood. This strategy is used within the GVNS I algorithm.

The *cyclic* strategy, adopted in GVNS II, visits the neighborhoods in a cyclic fashion. After applying a local search in the current neighborhood, the algorithm moves to the next one regardless of whether the search led to an improvement. Whenever a better solution is obtained, it becomes the new current solution, and the process continues until a full cycle is performed without improvement. This mechanism ensures a more uniform traversal of the neighborhood structures.

The *pipe* strategy, used in GVNS III, executes the local searches in a pipeline manner: once a neighborhood yields an improved solution, the algorithm immediately proceeds to the next neighborhood without restarting from the first one. This approach enables more rapid propagation of improvements across neighborhoods and may reduce the number of iterations needed to reach a locally optimal solution.

Across the three variants, the VND procedure serves as the core intensification component, systematically refining the shaken solution and exploiting complementary neighborhood structures. The resulting GVNS I, GVNS II, and GVNS III algorithms thus differ primarily in the way their respective VND components coordinate and sequence the neighborhood explorations.

5 Simulated Annealing

Simulated Annealing (SA) is a stochastic metaheuristic inspired by the controlled cooling process of heated metals, originally introduced by Kirkpatrick et al. [6]. Its probabilistic acceptance of non-improving moves allows the search to escape local optima, making it an effective strategy for complex combinatorial optimization problems. One of the earliest applications of SA in scheduling was presented by Osman and Potts [14] for the permutation flow shop problem.

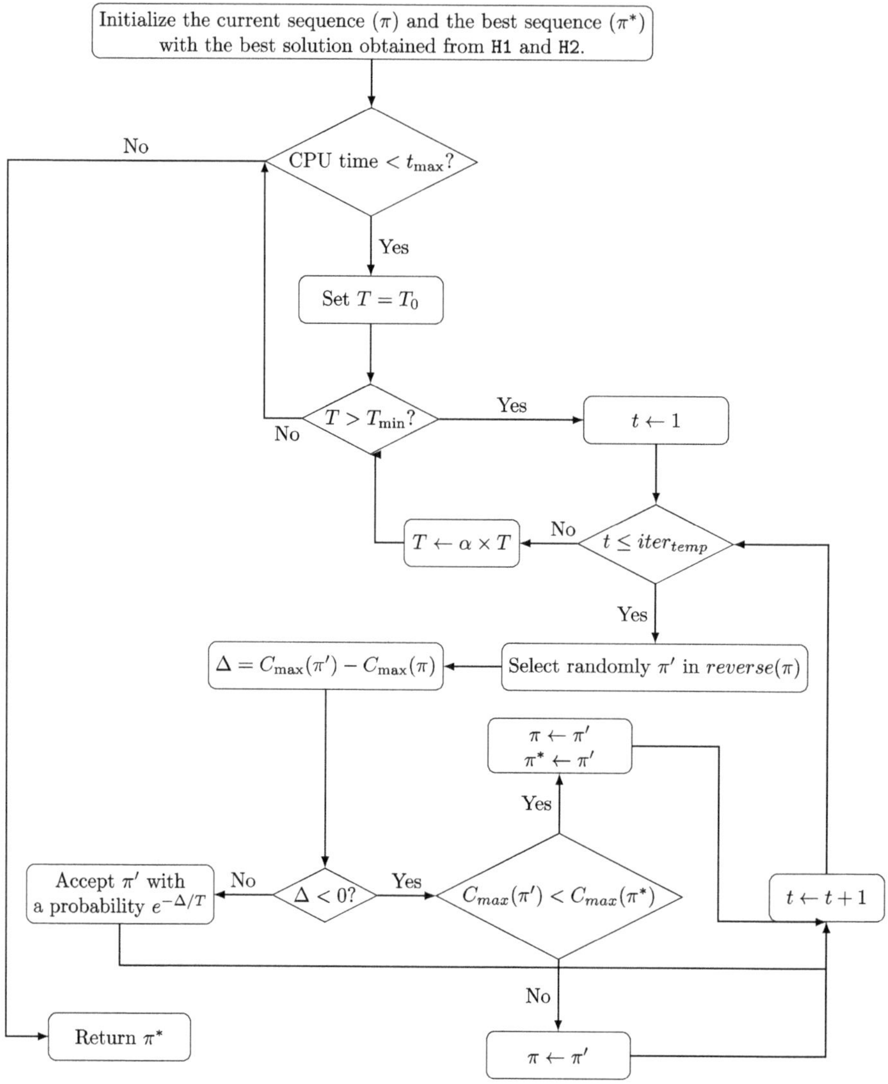

Fig. 4. Flowchart of the SA algorithm.

In this work, **SA** is adapted to the problem $F2(P_2, 1) \mid no\text{-}wait, S1 \mid C_{\max}$. The algorithm begins with an initial solution selected as the best among the two constructive heuristics **H1** and **H2**. At each iteration, a neighbor solution is generated using the *reverse* neighborhood. Improving moves are always accepted, while worsening moves may be accepted with probability $\exp(-\Delta/T)$, where Δ is the increase in makespan and T is the current temperature. For each temperature, this process is repeated $iter_{temp}$ times before the temperature is decreased geometrically according to $T \leftarrow \alpha T$, progressively shifting the search from diversification to intensification. This procedure is repeated until a time-based stopping condition is satisfied. The best solution obtained during the search is returned. The overall structure of the SA algorithm used in this work is illustrated in Fig. 4.

6   Computational Experiments

In this section, we conduct computational experiments to assess the effectiveness of the proposed methods, namely the MIP formulation, the constructive heuristics H1 and H2, the General Variable Neighborhood Search (GVNS) algorithm and the Simulated Annealing (SA) algorithm. Our benchmarking approach follows three main steps: (1) we evaluate the efficiency and limitations of the exact method for small instances for which the proposed metaheuristics also yield competitive solutions in shorter computation times; (2) we compare and analyze the performance of the heuristics and metaheuristics on larger instances (e.g., we measure the gaps with respect to the best-known solutions) (3) we evaluate the tightness of the proposed lower bounds on the small instances, for which provably optimal solutions are available (for the larger instances, where optimal solutions are not known, the reported gaps still provide a meaningful indication of the bounds' practical relevance and the heuristics' accuracy).

Sect. 6.1 describes the characteristics of the benchmark instances. Section 6.3 reports the performance of all proposed methods on small instances, while Sect. 6.4 focuses on the evaluation of heuristic and metaheuristic approaches on large instances.

The MIP formulation stops after a maximum runtime of 1 h. In contrast, GVNS and SA are executed until the solution obtained by the MIP model is reached for small instances. For large instances, the runtime of the metaheuristics is limited by a time cap defined as $t_{\max} = n \times 500$ milliseconds.

We solve the mixed-integer programming (MIP) model using CPLEX version 22.1.0 with default settings. All other methods are implemented in C++. All experiments are conducted on a personal computer equipped with an Intel(R) Core(TM) i7-12650H 2.30 GHz CPU and 16 GB of RAM, running Windows 11.

6.1   Instances Generation

Let n denote the number of jobs. Following the approach commonly adopted in the literature [8], we consider small instances with $n \in \{10, 15, 20, 30\}$ and large instances with $n \in \{50, 100, 200, 300\}$. Two sets of instances are generated. In the first set (Set 1), setup durations are drawn from a discrete uniform distribution over $[1, 50]$, while in the second set (Set 2), they are drawn over $[1, 100]$. In both sets, processing times for jobs are independently generated from a discrete uniform distribution over $[1, 100]$. For each value of n, 20 instances are created randomly.

6.2   Parameters Setting

The parameter values were chosen based on preliminary experiments. Specifically, for GVNS, we set the shaking parameter $k_{max} = 10$, while for SA, the initial temperature $T_0 = 10000$, the cooling factor $\alpha = 0.95$, the minimum temperature $T_{\min} = 10^{-3}$, and the number of iterations per temperature level $iter_{temp} = 100$. These values provided a good trade-off between exploration and computational efficiency.

6.3   Results for Small Instances

Table 2 presents the average results obtained for the small-sized instances across two sets. Each row reports the average over 20 instances. The first two columns indicate the instance set and the number of jobs n. Columns 3–6 show the makespan $C_{\max}$ and the relative gap $GAP_{LB*} = \frac{C_{\max}-LB^*}{LB^*} \times 100$ for the constructive heuristics H1 and H2. Columns 7–9 report the C_{max}, GAP_{LB*} and computational time (CPU) for the MIP model. The last four columns provide the CPU times for the metaheuristics (GVNS I--III and SA) needed to reach the best solution obtained by MIP. Bold values highlight the fastest metaheuristic for each instances set. Note that the CPU times for constructive heuristics are not reported, as they are negligible.

Table 2. Comparison of exact and heuristic methods on small instances.

Set	n	H1		H2		MIP			GVNS I	GVNS II	GVNS III	SA
		$C_{\max}$	GAP_{LB*}	$C_{\max}$	GAP_{LB*}	$C_{\max}$	GAP_{LB*}	CPU	CPU	CPU	CPU	CPU
Set 1	10	566.20	6.50	659.00	24.36	549.85*	3.06	1.69	**0.01**	**0.01**	**0.01**	0.25
	15	802.95	4.19	914.90	19.12	782.35*	1.29	577.45	0.59	**0.34**	0.95	1.60
	20	1036.95	3.05	1173.15	16.97	1023.90	1.62	3480.27	2.26	**1.38**	1.43	4.73
	30	1671.75	5.17	1820.10	14.72	1630.30	2.52	3600	4.47	**4.43**	4.56	5.22
Set 2	10	663.45	12.51	767.40	30.18	634.30*	7.40	1.24	0.01	0.01	**0.00**	0.04
	15	946.25	12.36	1073.65	27.78	902.85*	6.94	250.68	0.64	**0.33**	0.66	2.04
	20	1253.70	11.98	1427.25	27.56	1219.25	8.86	3348.50	2.22	**1.92**	2.36	4.50
	30	1931.95	11.84	2100.80	21.70	1880.50	8.89	3600	4.64	**3.53**	3.55	4.81
Avg		1109.15	8.45	1242.03	22.80	1077.91	5.07	1857.48	1.86	1.49	1.69	2.90

Based on the results presented in Table 2, the following observations can be made. Values reported as 0.00 for CPU time indicate extremely small runtimes (on the order of 10^{-4} to 10^{-5} s). The symbol "*" indicates that CPLEX reached the optimal solution for all instances in the corresponding category.

- The constructive heuristics H1 and H2 provide solutions almost instantly but with noticeably larger deviations from the best-known lower bounds. Overall, H1 systematically outperforms H2 in terms of both $C_{\max}$ and GAP_{LB*}. For example, in Set 1 with $n = 30$, H1 yields a makespan of 1671.75 (gap of 5.17%), while H2 obtains a significantly higher makespan of 1820.10 with a 14.72% gap. This trend is consistent across all instance sizes in both datasets.

- The MIP formulation provides the best overall solution quality and consistently returns optimal or near-optimal solutions for small instances ($n = 10$ and $n = 15$). However, it becomes computationally expensive for larger instances: for $n \geq 20$, most runs reach the time limit of $3600\,$s. Despite this limitation, the resulting solutions remain competitive, with gaps as low as 2.52% in Set 1 and 8.89% in Set 2 for $n = 30$.
- The metaheuristic methods GVNS I, GVNS II, and GVNS III demonstrate excellent computational efficiency, solving all instances in only a few seconds. Among them, GVNS II consistently achieves the best CPU performance. For example, in Set 2 with $n = 30$, GVNS II requires only $3.53\,$s, compared to $4.64\,$s for GVNS I and $3.55\,$s for GVNS III. The updated makespan values do not alter the relative ranking of these variants.
- The SA algorithm remains globally competitive but generally slower than the GVNS variants. For instance, in Set 1 with $n = 20$, SA requires $4.73\,$s, whereas GVNS II solves the same instance in only $1.38\,$s. Although SA produces acceptable solutions, it is less efficient computationally.

6.4 Results for Large Instances

Table 3 presents the results obtained for the large-sized instances. As with the small instances, each row reports the average results over 20 instances. To evaluate the performance of the proposed methods, we consider the following key metrics. The relative percentage deviation $RPD = \frac{C_{\max}(\text{heur}) - C_{\max}(\text{best})}{C_{\max}(\text{best})} \times 100$ which measures how much the makespan obtained by a given heuristic deviates from the best makespan achieved by all algorithms for a given instance. In addition, the optimality gap with respect to the lower bound (GAP_{LB*}) is also reported.

Table 3. Average RPD and GAP_{LB*} values for different algorithms on large instances.

Set	n	H1		H2		GVNS I		GVNS II		GVNS III		SA	
		RPD	GAP_{LB*}	RPD	GAP_{LB*}	RPD	GAP_{LB*}	RPD	GAP_{LB*}	RPD	GAP_{LB*}	RPD	GAP_{LB*}
Set 1	50	1.89	2.92	13.46	14.60	0.23	1.23	**0.03**	**1.03**	0.06	1.07	0.44	1.44
	100	1.11	2.45	11.56	13.04	0.06	1.39	**0.04**	**1.37**	0.06	1.38	0.15	1.48
	200	0.44	1.31	10.68	11.63	0.03	0.89	**0.01**	**0.87**	0.02	0.88	0.14	1.01
	300	0.76	1.94	10.28	11.57	**0.00**	**1.17**	**0.00**	1.17	0.02	1.18	0.32	1.49
Set 2	50	3.91	10.31	17.25	24.51	0.52	6.72	0.37	6.55	**0.23**	**6.41**	1.46	7.71
	100	3.71	9.09	14.49	20.43	0.55	5.77	**0.18**	**5.38**	0.25	5.46	1.81	7.10
	200	4.02	9.88	14.97	21.44	0.52	6.18	**0.15**	**5.79**	0.22	5.87	3.33	9.15
	300	3.43	8.53	13.43	19.02	0.46	5.40	**0.11**	**5.04**	0.13	5.06	4.05	9.18
Avg		2.41	5.80	13.27	17.03	0.30	3.59	0.11	3.40	0.12	3.41	1.46	4.82

From the results in Table 3, several conclusions can be drawn regarding the performance of the algorithms on large instances:

- The constructive heuristics H1 and H2 continue to provide the weakest results, with consistently high RPD and GAP_{LB*} values. In particular, H2 shows very large deviations from the best solutions, confirming that these heuristics are mainly useful for generating initial sequences that must be improved by more advanced methods.
- Across all instance sizes, the GVNS variants clearly dominate the other methods. In Set 1, both GVNS I and GVNS II achieve very small deviations, with RPD values often close to zero. GVNS II slightly outperforms GVNS I in several cases, obtaining the smallest GAP_{LB*} values.
- For Set 2, which contains more challenging instances, GVNS II remains the most competitive algorithm. It systematically achieves the smallest RPD and GAP_{LB*} values, followed closely by GVNS III. These results highlight the robustness of GVNS II when the instance difficulty increases.
- The Simulated Annealing (SA) algorithm provides reasonable results but is consistently outperformed by all GVNS variants, particularly for larger instances where both RPD and GAP_{LB*} become significantly higher.
- Overall, the GVNS algorithms exhibit strong and stable performance as the problem size increases. GVNS II generally gives the best results on both sets, while the constructive heuristics (H1 and H2) remain far from optimal, with large deviations that confirm their limited effectiveness for high-quality solution generation.

6.5   Statistical Analysis

All statistical analyses and plots were performed using RStudio (version 4.3.2) with base R packages. Figure 5 presents the boxplot of the Relative Percentage Deviation (RPD) values obtained by the evaluated algorithms on instances of size $n = 50$. This figure provides a concise visual comparison of the central tendency and variability of RPD for each method.

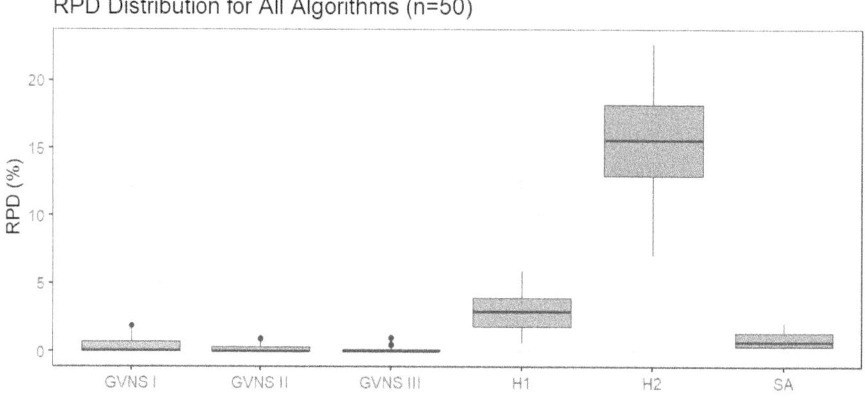

Fig. 5. Boxplots of RPD values for all algorithms on instances with $n = 50$.

From the boxplot, three main observations can be made. First, the GVNS variants achieve lower median RPD values and substantially narrower interquartile ranges than the constructive heuristics (H1, H2) and Simulated Annealing (SA), indicating both superior average solution quality and lower variability. Second, the constructive heuristics, especially H2, exhibit larger RPDs and more dispersed distributions, confirming that these fast construction rules produce solutions that are further from the best-found values. Third, SA demonstrates intermediate behavior: its medians and spreads are higher than those of the GVNS variants but generally lower than those of the constructive heuristics.

For completeness, a full set of statistical tests (Shapiro–Wilk normality tests, Kruskal–Wallis tests, and pairwise Wilcoxon rank–sum comparisons with correction) was performed for all instance sizes to assess the significance of observed differences. However, these detailed results for other instance sizes ($n = 100, 200, 300$) are omitted here for brevity, as they consistently confirm the visual impressions: GVNS variants outperform the constructive heuristics and SA, while differences among GVNS variants are mostly not statistically significant at the 0.05 level.

7   Conclusions and Future Directions

This paper studied the two-stage hybrid flow shop with a single server at the first stage, two parallel machines in stage one, and a single machine in stage two, under no-wait and server constraints. A mixed-integer programming (MIP) model was proposed for small instances, while two constructive heuristics and metaheuristics, including General Variable Neighborhood Search (GVNS) and Simulated Annealing (SA), were developed for larger instances.

Computational results showed that the MIP model solves small instances optimally, heuristics provide good initial solutions, and GVNS variants outperform SA, achieving the lowest average relative percentage deviations and gaps.

Limitations include the MIP's scalability and varying performance of heuristics and metaheuristics for large sized instances. Future work could consider relaxing the no-wait constraint, multiple servers, sequence-dependent setups, limited buffers, or machine breakdowns.

Availability of Data and Materials. The data and Results that support the findings of this study are publicly accessible at: http://dx.doi.org/10.17632/tn7tgd8c6s.1.

References

1. Allahverdi, A.: A survey of scheduling problems with no-wait in process. Eur. J. Oper. Res. **255**(3), 665–686 (2016)
2. Balas, E.: On the facial structure of scheduling polyhedra. In: Cottle, R.W. (eds.) Mathematical Programming Essays in Honor of George B. Dantzig Part I, pp. 179–218. Springer, Cham (2009). https://doi.org/10.1007/BFb0121051

3. Benmansour, R., Sifaleras, A.: Scheduling in parallel machines with two servers: the restrictive case. In: Mladenovic, N., Sleptchenko, A., Sifaleras, A., Omar, M. (eds.) ICVNS 2021. LNCS, vol. 12559, pp. 71–82. Springer, Cham (2021). https://doi.org/10.1007/978-3-030-69625-2_6
4. Chang, J., Yan, W., Shao, H.: Scheduling a two-stage no-wait hybrid flowshop with separated setup and removal times. In: Proceedings of the 2004 American Control Conference, vol. 2, pp. 1412–1416. IEEE (2004)
5. Graham, R.L., Lawler, E.L., Lenstra, J.K., Rinnooy Kan, A.H.G.:. Optimization and approximation in deterministic sequencing and scheduling: a survey. In: Annals of Discrete Mathematics, vol. 5, pp. 287–326. Elsevier (1979)
6. Kirkpatrick, S., Daniel Gelatt Jr., C., Vecchi, M.P.: Optimization by simulated annealing. Science $220(4598)$, 671–680 (1983)
7. Krim, H., Benmansour, R., Duvivier, D., Artiba, A.: A variable neighborhood search algorithm for solving the single machine scheduling problem with periodic maintenance. RAIRO-Oper. Res. $53(1)$, 289–302 (2019)
8. Kurt, A.: Integrating sequence-dependent setup times and blocking in hybrid flow shop scheduling to minimize total tardiness. Int. J. Ind. Eng. Comput. $16(1)$, 147–158 (2025)
9. Liu, M., Sun, Z., Zhang, X., Chu, F.: A two-stage no-wait hybrid flow-shop model for the flight departure scheduling in a multi-airport system. In: 2017 IEEE 14th International Conference on Networking, Sensing and Control (ICNSC), pp. 495–500. IEEE (2017)
10. Medouar, F., Elidrissi, A., Benmansour, R., Gupta, J.N.D.: No-wait two-machine permutation flow shop scheduling problem with a single server and separable setup times to minimize total tardiness. In: 2023 9th International Conference on Control, Decision and Information Technologies (CoDIT), pp. 626–631. IEEE (2023)
11. Mladenović, N., Hansen, P.: Variable neighborhood search. Comput. Oper. Res. $24(11)$, 1097–1100 (1997)
12. Moradinasab, N., Shafaei, R., Rabiee, M., Ramezani, P.: No-wait two stage hybrid flow shop scheduling with genetic and adaptive imperialist competitive algorithms. J. Exp. Theor. Artif. Intell. $25(2)$, 207–225 (2013)
13. Nawaz, M., Enscore, E.E., Jr., Ham, I.: A heuristic algorithm for the m-machine, n-job flow-shop sequencing problem. Omega $11(1)$, 91–95 (1983)
14. Osman, I.H., Potts, C.N.: Simulated annealing for permutation flow-shop scheduling. Omega $17(6)$, 551–557 (1989)
15. Ling-Huey, S., Lee, Y.-Y.: The two-machine flowshop no-wait scheduling problem with a single server to minimize the total completion time. Comput. Oper. Res. $35(9)$, 2952–2963 (2008)
16. Utama, D.M., Umamy, S.Z., Al-Imron, C.N.: No-wait flow shop scheduling problem: a systematic literature review and bibliometric analysis. RAIRO-Oper. Res. $58(2)$, 1281–1313 (2024)
17. Wang, S., Liu, M.: A genetic algorithm for two-stage no-wait hybrid flow shop scheduling problem. Comput. Oper. Res. $40(4)$, 1064–1075 (2013)
18. Wang, S., Wang, X., Li, Yu.: Two-stage no-wait hybrid flow-shop scheduling with sequence-dependent setup times. Int. J. Syst. Sci. Oper. Logist. $7(3)$, 291–307 (2020)
19. Xuan, H., Zheng, Q., Li, B., Wang, X.: A novel genetic simulated annealing algorithm for no-wait hybrid flowshop problem with unrelated parallel machines. ISIJ Int. $61(1)$, 258–268 (2021)
20. Zhong, W., Shi, Y.: Two-stage no-wait hybrid flowshop scheduling with inter-stage flexibility. J. Comb. Optim. $35(1)$, 108–125 (2018)

Solving the Integrated Production and Routing Problem Using a Skewed Hybrid Variable Neighborhood Search Algorithm

Mário Leite, Telmo Pinto[✉], and Cláudio Alves

Centro ALGORITMI/LASI, Universidade do Minho, Campus de Gualtar, 4710-057 Braga, Portugal
{mario.leite,telmo,claudio}@dps.uminho.pt

Abstract. This paper addresses the Integrated Production Routing Problem (IPRP), which requires the joint coordination of lot-sizing and vehicle routing decisions over a finite planning horizon. The problem accounts for several real-world features, including multiple products with different items, weights and sizes, sequence-dependent setup times and costs, limited production capacity, and safety stock requirements. For distribution operations, a heterogeneous fleet must serve customers with multiple time windows and strict deadlines, with routes possibly spanning several periods. The objective is to determine an integrated production and routing plan that minimizes setup changeover, inventory holding, and transportation costs. For this purpose, a Skewed Hybrid Variable Neighborhood Search (SHVNS) algorithm is proposed, relying on neighborhood structures specifically designed to modify routing decisions, while production planning is obtained from a mixed-integer programming model. Extensive computational experiments on benchmark instances demonstrate the effectiveness of the method in solving large-scale instances.

Keywords: Production and Routing · Optimization · Hybrid Algorithms

1 Introduction

The continuous progress observed in the field of combinatorial optimization has led researchers to tackle increasingly realistic and complex scenarios. A good example is the attempt to solve simultaneously interdependent problems in an integrated way, by considering the interactions between them. The problem addressed in this paper belongs to this category. It was first defined based on a real-world industrial case and involves the joint planning of production activities and the corresponding distribution operations required to deliver the produced items to customers. Formally, the problem combines a lot-sizing subproblem to

S. Cavero et al. (Eds.): ICVNS 2025, LNCS 16256, pp. 79–94, 2026.
https://doi.org/10.1007/978-3-032-19582-1_6

determine the quantities to produce and when to produce them, with a vehicle routing subproblem. In the literature, these problems are also referred to as rich optimization problems.

In recent years, there has been a growing interest in solving lot-sizing problems that explicitly consider their interaction with distribution operations. A survey emphasizing the benefits of solving these problems simultaneously was recently published by Darvish et al. in [1]. Further contributions can be found in [2,9,10], for example. In [9], the authors explore a multi-item dynamic lot-sizing problem combined with distribution that allows item fragmentation and different transportation modes. The computational results they obtained show that optimizing production and distribution simultaneously leads to a clear reduction in the global costs. An integrated lot-sizing and vehicle routing problem arising in the context of the distribution of perishable goods was addressed by Yağmur et al. [10]. The authors explore a model that includes different real-world constraints. Their study shows how the combined optimization of production and distribution can improve both cost-efficiency and sustainability in supply chains involving perishable products. Gruson et al. [2] study an integrated three-level lot-sizing and distribution problem that allows demand splitting and partial deliveries. Their results show that integrated planning across multiple supply chain levels leads to more efficient and flexible operations compared to the sequential approach.

In this paper, we explore a variant of the so-called Production Routing Problem (PRP), as defined in [6], which is inspired by a real case from a small make-to-order furniture company. The problem considers the possibility of having different products composed of different components with varying weights and sizes, sequence-dependent setup times and costs, a heterogeneous vehicle fleet, multi-period routing, safety stock requirements for each product component, and customers with multiple delivery time windows and deadlines. The objective is to jointly define the production and distribution plans in an integrated manner, by computing the production quantities and the sequencing plan of each item for every period, while simultaneously designing the routing plan specifying when and how customers will be served. The global objective is to minimize the total cost resulting from both production and distribution operations. For this purpose, Miranda et al. [7] developed a Mixed-Integer Programming (MIP) formulation that models the interdependencies between the scheduling and the routing decisions. A heuristic is also described to tackle the medium and large scale instances. The methods proposed in [6] follow a sequential approach, where the lot-sizing subproblem is solved first, and the resulting production plan is then used to determine the distribution routes. Several papers in the literature report on results for this problem using a similar approach [5,8].

In this paper, we explore a hybrid approach to solve the rich Production Routing Problem proposed by Miranda et al. [6]. Our method relies on a Skewed Hybrid Variable Neighborhood Search (SHVNS) metaheuristic and on new neighborhood structures. Extensive computational experiments were conducted on a set of benchmark instances from the literature. The results show clear improvements over existing heuristics, particularly for large-scale instances. Moreover, these improvements tend to become even more significant as the size of the instances increases.

The remainder of the paper is organized as follows. In Sect. 2, the specific variant of the PRP addressed in this work is formally defined. In Sect. 3, we present our hybrid metaheuristic based on Skewed Variable Neighborhood Search. In Sect. 4, we describe the benchmark instances used in our computational experiments and report the results obtained. Finally, Sect. 5 summarizes the main contributions of this paper and suggests potential directions for future research.

2   The Integrated Production Routing Problem

As mentioned above, this paper explores a variant of the Integrated Production Routing Problem (IPRP) defined in the context of a make-to-order furniture manufacturer, which produces different components/items used in the assembly of different final products. The problem involves determining the quantities of components to be produced in each time period and defining their production sequence, ensuring that the necessary items are available to supply customers through delivery routes. The overall objective is to satisfy demand while minimizing the total cost of production and distribution, taking into account the assumptions described below.

The characteristics of the problem, including its constraints, are summarized below.

- **Time periods:** the planning horizon is divided into equal time periods. In each period, the production facility operates for a given duration and can produce multiple items. However, performing a changeover of the item to be produced entails a setup cost and time. Setups are continuous and may extend across periods.
- **Production:** items produced in a given period cannot be shipped during the same period; they are available for delivery only in the subsequent one.
- **Inventory:** initial and minimum inventory levels of the items are considered, and a holding cost is applied to each unit stored at the facility before shipment, with the cost determined by the two-dimensional size of the item (width and length).
- **Distribution:** the distribution network is represented by a complete directed graph, in which the nodes correspond to all locations involved and the arcs denote the possible travel connections between them. Each connection has an associated travel time and cost.
- **Time windows:** each location has a specific working time window for every period. For customer locations, the vehicle must begin service within that time window or wait if it arrives early. At each customer location, the service must start at any moment within the assigned time window.
- **Vehicles:** the fleet is heterogeneous, with each vehicle having a specific capacity expressed in units of weight.

- **Customers:** each customer must be visited exactly once to meet its demand for each product within a specified delivery deadline, and the service time is proportional to the total weight of the requested products, with the loading and unloading time determined by the amount of weight handled.
- **Routes:** deliveries are organized into a set of routes, whose total number is bounded in advance. Routes may either extend over several periods or be restricted to a single one, but they always start and end at the depot after visiting their assigned customers.
- **Loading:** between successive routes, each vehicle is loaded at the depot, according to the demand of its assigned customers. The loading time equals the sum of their service times and must be completed within the depot time window for that period. The latest moment at which loading can start is determined by subtracting, from the end of the depot time window, the time needed to handle the total demand of all customers in that route.

For the sake of clarity, we present in Table 1 and Table 2 the terminology and notation used throughout this paper.

Table 1. List of symbols and their meaning (Part I)

Symbol	Meaning
$\mathcal{T}$	Set of time periods (time horizon)
K_t	Available production time in period $t \in \mathcal{T}$
$\mathcal{P}$	Set of final products
φ_p	Weight of one unit of product $p \in \mathcal{P}$
$\mathcal{C}$	Set of components (items)
η_{ap}	Units of component $a \in \mathcal{C}$ required to produce one unit of product $p \in \mathcal{P}$
$\mathcal{F}_p \subseteq \mathcal{C}$	Set of components required to assemble product $p \in \mathcal{P}$
ρ_a	Time to produce one unit of item $a \in \mathcal{C}$
$\hat{c}_{ab}$	Setup cost for switching from a to b $(a, b \in \mathcal{C})$
ς_{ab}	Setup time for switching from a to b $(a, b \in \mathcal{C})$
I_{a0}	Initial inventory level of item $a \in \mathcal{C}$
$I_a^{\min}$	Minimum inventory level of item $a \in \mathcal{C}$
h_a	Holding cost per unit of item $a \in \mathcal{C}$ per period
W_a	Width of item $a \in \mathcal{C}$
L_a	Length of item $a \in \mathcal{C}$
$\mathcal{V}$	Set of vehicles (fleet)
θ_v	Capacity of vehicle $v \in \mathcal{V}$ (weight units)
R	Upper bound on the number of routes
$\bar{\mathcal{R}}$	Set of customers assigned to a given route

Table 2. List of symbols and their meaning (Part II)

Symbol	Meaning
$\mathcal{N}$	Set of nodes
$\mathcal{A}$	Set of directed arcs
$\mathcal{G} = (\mathcal{N}, \mathcal{A})$	Complete directed graph of the network
$\bar{\mathcal{C}} = \mathcal{N} \setminus \{0, n+1\}$	Set of customers
$0, n+1$	Nodes representing the depot (0 for departure, $n+1$ for return)
$(i, j) \in \mathcal{A}$	Arc from node i to node j, $i, j \in \mathcal{N}$
τ_{ij}	Travel time on arc (i, j), $i, j \in \mathcal{N}$
c_{ij}	Travel cost on arc (i, j), $i, j \in \mathcal{N}$
$[\delta_{it}, \bar{\delta}_{it}]$	Time window for node $i \in \mathcal{N}$ in period $t \in \mathcal{T}$
d_{pi}	Demand of customer $i \in \bar{\mathcal{C}}$ for product $p \in \mathcal{P}$
Δ_i	Delivery deadline for customer $i \in \bar{\mathcal{C}}$
λ	Time to load/unload one unit of weight
s_i	Service time at customer $i \in \bar{\mathcal{C}}$ given by $\lambda \sum_{p \in \mathcal{P}} \varphi_p d_{pi}$

The following example illustrates some of the main features of a small size instance.

Example 1. Consider a small instance with a planning horizon divided into $|\mathcal{T}| = 5$ time periods, and a production capacity in each period equal to $K_t = 250$ time units. There are two final products ($\mathcal{P} = \{1, 2\}$), each composed of two types of items ($\mathcal{C} = \{a, b\}$), with equal processing times for both items ($\rho_a = \rho_b = 1$). The number of item units required to produce one unit of product 1 is $\eta_{a1} = 4$ and $\eta_{b1} = 1$; for product 2, the requirements are $\eta_{a2} = 2$ and $\eta_{b2} = 1$.

Table 3 presents the final product demand d_{pi} for each of the five customers ($n = 5$). The corresponding item demand for each customer is shown in parentheses. For instance, if customer 1 requests 10 units of each product, the required number of units of item a will be $\eta_{a1} \times d_{1,1} + \eta_{a2} \times d_{2,1} = 4 \times 10 + 2 \times 10 = 60$ units, and for item b: $\eta_{b1} \times d_{1,1} + \eta_{b2} \times d_{2,1} = 1 \times 10 + 1 \times 10 = 20$ units.

Table 3. Customers demand

Customer $i \in \bar{\mathcal{C}}$	1	2	3	4	5
Product 1 (Item#a)	10 (60)	20 (120)	10 (80)	5 (70)	10 (80)
Product 2 (Item#b)	10 (20)	10 (40)	20 (30)	25 (30)	20 (30)

The setup changes between items require $\varsigma_{ab} = 10$ and $\varsigma_{ba} = 30$ units of time. For the sake of simplicity, both the minimum and the initial inventory levels are assumed to be zero, i.e., $I_a^{\min} = I_{a0} = 0$.

In Fig. 1, a possible solution for this small instance of the PRP is depicted.

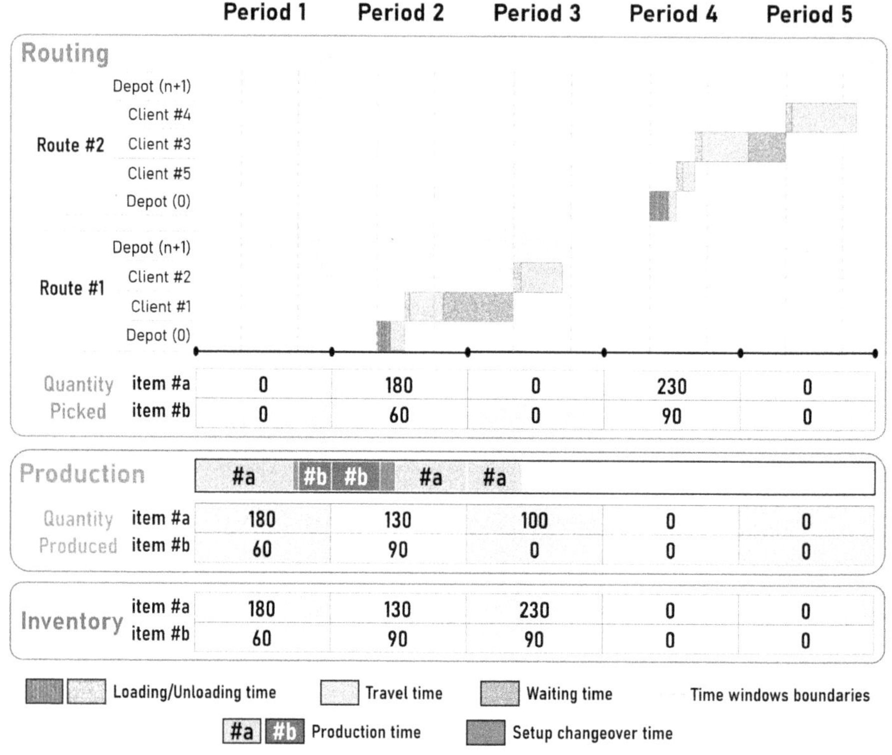

Fig. 1. Illustrative production and routing plan (Example 1)

In this plan, the entire production capacity is used in period 1 ($K_1 = 250$). Production starts with 180 units of item a, followed by a setup change requiring ς_{ab} time units, and then 60 units of item b are produced. The total time consumption is $180 \times \rho_a + \varsigma_{ab} + 60 \times \rho_b = 250$. In period 2, in addition to new production, route #1 is executed, covering the full demand of the customers it visits (customers 1 and 2). Route #2 begins in period 4 and visits the remaining customers. The inventory (or stock level) in a period $t \in \mathcal{T}$ is defined as the difference between the total quantities produced and distributed until period t.

The total cost (z) associated with this solution is given by the following expression:

$$z = \hat{c}_{ab} + \hat{c}_{ba} + h_a \times (180 + 130 + 230) + h_b \times (60 + 90 + 90) + c_{01} + c_{12} + c_{26} + c_{05} + c_{53} + c_{54} + c_{46}.$$

The first two terms correspond to the setup changeover costs. The next two terms represent inventory holding costs. Finally, the remaining terms account for routing costs, which include the travel costs between nodes.

3   A Skewed Hybrid VNS Algorithm

To solve the IPRP described in the previous section, we developed a skewed hybrid variable neighborhood search (SHVNS) algorithm with new neighborhood structures that proved to outperform the current state-of-the-art approaches for this problem, particularly on large-scale instances. In this section, we describe the main characteristics of our solution approach. We emphasize the key aspects that distinguish it from the heuristic proposed in [4], in particular the skewed variable neighborhood search component on which our algorithm relies primarily.

3.1   Initialization

Our algorithm starts by computing an initial solution for the IPRP by using the mathematical programming based heuristic proposed in [4]. The method assigns each generated route to the vehicle with the highest available capacity, inserting customers according to a nearest-neighbor rule as long as the capacity and time-window constraints are met; otherwise the next feasible customer is selected. When no further insertions are possible, a new route is created or postponed until a vehicle becomes available. After all customers have been assigned, a mixed-integer programming (MIP) model is used to determine the optimal production quantities and schedule for each period, based on the previously computed routes. The model aims to minimize the total cost, comprising inventory holding and production setup costs, while scheduling items over discrete time periods. It incorporates constraints that limit the available time for production and setups, addresses changeover and sequencing requirements, satisfies delivery lead times, and ensures that inventory levels comply with the required thresholds. Given the size of the problems and the fact that it does not depend on the number of customers, the MIP model can be solved in a very short computational time (typically a few milliseconds). This model is used not only to generate an initial solution but also repeatedly during the SHVNS search process, as discussed later.

3.2   Neighborhood Structures

To diversify the exploration of the solution space in the Variable Neighborhood Search part of our algorithm, we use five different neighborhood structures (NS_k, $k \in \{1, 2, 3, 4, 5\}$) based on the definition of the vehicle routes. The aim is to determine good quality solutions by exploring the corresponding space of vehicle routes, using the production plan computed exactly by solving the MIP mentioned above.

- NS_1: neighbors are obtained by modifying the departure period of a route;
- NS_2: neighbors are generated by moving a customer from one route to another;
- NS_3: neighbors are generated by removing a customer from a multi-customer route and inserting it into a new single-customer route;
- NS_4: neighbors are obtained by swapping two customers within the same route;

– NS_5: neighbors are generated by exchanging two customers from different routes.

The first neighborhood structure (NS_1) was introduced to explore simultaneously the production and the routing plan. Note that, although changes in the departure times do not affect the routing costs, they may significantly influence the production schedule. Advancing departures prioritizes urgent demand, whereas delaying them supports batch consolidation but may increase inventory levels. Similarly, but in the opposite direction, the neighborhood structure (NS_4) does not affect the production plan cost, since the departure time remains unchanged due to the intra-route nature of the swaps. However, changing the order in which customers are visited within the same route can affect the routing cost.

The remaining neighborhood structures (from NS_2 to NS_5) imply a change in the order of the customer visits that impacts the transportation costs. In NS_2 and NS_3, the relocation of a single customer (either to another existing route or to a new route) is intended to provide flexibility to merge or split routes.

The neighborhood structure NS_5 explores the simultaneous exchange of two customers from different routes as a way to search for solution spaces where vehicle capacity may become a constraint. In other words, it aims to overcome the limitation that arises when a vehicle is close to its capacity and a new customer cannot be inserted into the route. By swapping two customers, the demand of the customer removed from one route, combined with the available capacity, may allow the insertion of the new customer.

3.3 Skewed Variable Neighborhood Search

The proposed Skewed Hybrid Variable Neighborhood Search (SHVNS) algorithm follows the general principles for skewed VNS described in [3]. Our approach is summarized in Algorithm 1.

At each iteration, the algorithm applies a shaking function based on the neighborhood structure k, intensified by a parameter m (performing m consecutive moves), in order to diversify the routing plans explored (line 7). This is followed by a local search procedure based on the NS_2 neighborhood structure, which has proven more effective than the other structures, with the aim of improving the quality of the current solution (line 8). For the resulting set of routes $(S'_{partial})$, the optimal production plan is then obtained by solving the MIP mentioned in the previous section (line 9).

As a skewed variant of the VNS, our method allows the exploration of potentially worse solutions, provided that they are sufficiently different from the best known solution (line 15) according to a distance function $\rho(S, S')$. The objective is to escape local optima and explore other promising regions of the solution space.

Let $\sigma(S)$ denote a score function, where td_{at} represents the total quantity of items $a \in \mathcal{C}$ loaded onto the vehicles in period $t \in \mathcal{T}$, as follows

$$\sigma(S) = \sum_{t \in \mathcal{T}} \sum_{a \in \mathcal{C}} t \times td_{at} . \tag{1}$$

A lower score indicates that a larger number of items is loaded in the earlier periods of the planning horizon, whereas a higher score indicates that more items are loaded in periods closer to the end of the horizon. For instance, the solution in Example (1) would have a score $\sigma(S_{example}) = 2 \times 180 + 2 \times 60 + 4 \times 230 + 4 \times 90 = 1760$. Considering the boundary cases in which all 560 items are assigned to the routes in the first period, the score would be 1×560; if all items were assigned to the routes in the last period, the score would be $5 \times 560 = 2800$. Hence, a possible distance function $\rho(S, S')$ between two solutions S and S' can correspond to the normalized absolute difference between their scores, $\sigma(S)$ and $\sigma(S')$, respectively, with $\rho(S, S') \in [0, 1]$, as follows:

$$\rho(S, S') = \frac{|\sigma(S) - \sigma(S')|}{(|\mathcal{T}| - 1) \times \sum_{a \in \mathcal{C}} \sum_{p \in \mathcal{P}} \sum_{i \in \bar{\mathcal{C}}} \eta_{ap} d_{pi}} \ . \tag{2}$$

A given solution S' is accepted if $z(S') < (1 + \alpha\rho(S,S')) \times z(S)$, with $\alpha \in [0, 1]$ (line 15). If this is the case, the neighborhood index is reset (line 17). Otherwise the neighborhood index and the value of m (perturbation intensity) are incremented in order to explore new solution spaces (lines 19–23).

Algorithm 1. Skewed Hybrid VNS Algorithm

1: **function** SKEWED-HYBRID-VNS(S, k_{max}, t_{max}, m_{max}, NS, α)
2:      $t_{start} := \text{CPUTime}()$; $t := 0$; $m := 1$;
3:      **while** $t \leq t_{max}$ **do**
4:          $k := 1$;
5:          $S_{best} := S$;
6:          **while** $k \leq k_{max}$ **do**
7:             $S_{partial} := Shake(S, NS_k, m)$;
8:             $S'_{partial} := LocalSearch(S_{partial}, NS_2)$;
9:             $S' := IPModel_{production}(S'_{partial})$;
10:            **if** $z(S') < z(S_{best})$ **then**
11:               $S_{best} := S'$;
12:            $t := CPUTime() - t_{start}$;
13:            **if** $t > t_{max}$ **then**
14:               $\alpha := 0$;
15:            **if** $z(S') < (1 + \alpha\rho(S, S'))z(S)$ **then**
16:               $S := S'$;
17:               $k := 1$;
18:            **else**
19:               $k := k + 1$;
20:               **if** ($m = m_{max}$) **then**
21:                  $m := 1$;
22:               **else**
23:                  $m := m + 1$;
24:      $t := CPUTime() - t_{start}$;
25:      $S := S_{best}$;

---

4 Computational Experiments

4.1 Instances

The performance of the algorithm was evaluated through computational experiments on real-world instances from the literature [6]. In these instances, the planning horizon was divided into eight time periods ($|T| = 8$), and the customer time windows were set according to their working hours. The inventory costs reflect the heterogeneity in item dimensions (width and length).

From the set of instances used in [6], we chose the instances with the largest numbers of customers, namely those with $n \in \{30, 40, 50\}$. An instance in these sets is obtained by varying the following parameters: the number of final products, $\mathcal{P} \in \{3, 5\}$, the number of items, $\mathcal{C} \in \{3, 5\}$, and the fleet size $\mathcal{V} \in \{2, 3\}$. Five distinct instances were generated for each parameter configuration, resulting in a total of 120 instances.

4.2 Results and Discussion

Because of the stochastic nature of the VNS-based approaches, each instance was solved independently five times on a PC with an Intel Core i7-8565U (8th generation) processor and 16 GB of RAM. The algorithm was implemented in Java, and IBM ILOG CPLEX 12.10.0 with Concert Technology serving as the mathematical programming solver.

The parameter settings of Algorithm 1 were defined based on an empirical analysis carried out through multiple computational experiments using different algorithmic configurations. Accordingly, the number of neighborhood structures was set to $k_{max} = 5$, and the execution time limit was fixed at $t_{max} = 150$ seconds. The parameter controlling the maximum number of consecutive perturbations (m_{max}) was limited to 10 moves, while the parameter α was set to 0.1, allowing solutions up to 10% worse than the best solution found during the search. The performance of our algorithm was evaluated by using the best value obtained over five runs.

To evaluate the relative performance of our algorithm, we compared our results with the state-of-the-art approaches described in the literature for the IPRP, namely the exact solution of the problem using a MIP formulation, the decomposition heuristic H presented in [6], and the hybrid VNS algorithm (H-VNS) described in [4]. In [6], the authors demonstrate clearly the limitations of an exact approach based on solving a MIP model for the complete IPRP. Even with 24 h of computation, the MIP model was able to solve only four out of 120 instances with 30 or more customers. These results justify the development of heuristics for this problem.

The H-VNS approach proposed in [4] consists of three explicit phases. The first is an initialization phase, in which the initial solution is generated. The second is a routing plan improvement phase, where a basic VNS procedure is applied using routing-based neighborhood structures, with the goal of reducing the routing cost. The final phase uses a hybrid strategy that integrates a MIP model restricted to the production plan into another VNS scheme, in which the traditional local search step is replaced by this restricted MIP model. The results presented in [4] show that H-VNS clearly outperforms the MIP formulation and the heuristic H described in [6]. However, as the size of the instances increases, we observe a clear degradation in the performance of H-VNS. These limitations motivated the development of the SHVNS.

In our experiments, the time limit for the execution of SHVNS was set to 150 s. To ensure a fair comparison with the current state-of-the-art algorithm for the IPRP (H-VNS), we repeated the tests using H-VNS under the exact same conditions as those used to test SHVNS. These tests of H-VNS were run with a time limit of 150 s on the same machine.

When comparing the results obtained from two approaches, say A and B, the relative difference between the evaluated metrics is computed as $((value^A - value^B)/value^B) \times 100$. In the following tables, a negative value indicates that A outperforms B in this metric. In our tables, the values in the columns Dev'', $DevToH'$, and $DevToHVNS$ are shown in bold whenever SHVNS performs at least as well as the other approach.

The computational results are summarized in Tables 4, 5 and 6. The meaning of each column is as follows:

- *Inst*: ordinal number of the instance;
- $|\mathcal{P}|$: number of final products;
- $|\mathcal{C}|$: number of items;
- $|\mathcal{V}|$: number of vehicles;
- *Best* and *Best'*: best solution found by H-VNS described in [4] and SHVNS approaches, respectively;
- *AvgSol*: average solution value computed over five independent runs;
- σ: standard deviation based on the five independent runs;
- $CV(\%)$: coefficient of variation, also known as the relative standard deviation (RSD), computed as $\sigma/AvgSol \times 100$;
- CPU: total computing time (in seconds);
- $CPU*$: time to obtain the best feasible solution (in seconds);
- $Dev(\%)$ and $Dev'(\%)$: relative deviation between *Best* and UB^{MIP}, *Best'* and UB^{MIP}, respectively. The term UB^{MIP} refers to the best feasible solution found by the MIP method described in [6] (or " $-$ " if no feasible solution was found by the MIP);
- $DevToH(\%)$ and $DevToH'(\%)$: relative deviation between *Best* (solution of the H-VNS) and UB^H, and between *Best'* (solution by SHVNS) and UB^H, respectively, where UB^H denotes the best solution obtained by the heuristic H described in [6];
- $DevToHVNS(\%)$: relative deviation between *Best'* (SHVNS) and *Best* (H-VNS);

From the results in Table 4, it is clear that H-VNS [4] performs better than the heuristic approach H of [6]. On the other hand, its performance deteriorates for the most challenging instances, with an average relative deviation from H ($DevToH$) of 0.59 and 1.15 for the sets of instances with 40 and 50 customers, respectively.

Table 4. Computational results for instances with 30 customers

Inst.	$|\mathcal{P}|$	$|\mathcal{C}|$	$|\mathcal{V}|$	H-VNS					SHVNS								
				Best	*CPU*	*CPU**	*Dev*	*DevToH*	*Best'*	*AvgSol*	σ	*CV*	*CPU*	*CPU**	*Dev'*	*DevToH'*	*DevToHVNS*
1	3	3	2	6799.2	152.2	22.5	–	−0.99	6714.7	6726.7	21.6	0.32	150.2	115.9	–	**−2.22**	**−1.24**
2	3	3	2	7634.6	152.2	9.4	−9.06	−1.81	7580.1	7653.9	64.3	0.84	150.1	60.5	**−9.71**	**−2.52**	**−0.71**
3	3	3	2	7913.7	152.2	56.1	–	1.05	7802.1	7922.7	80.1	1.01	150.2	83.1	–	**−0.38**	**−1.41**
4	3	3	2	7191.0	152.4	4.7	–	−0.46	7262.0	7272.5	6.8	0.09	150.2	88.5	–	0.52	0.99
5	3	3	2	6331.0	152.3	17.7	–	−2.01	6090.3	6263.7	86.8	1.39	150.1	52.4	–	**−5.73**	**−3.80**
6	3	3	3	6101.9	152.2	6.1	−28.81	−0.84	6057.5	6144.4	60.6	0.99	150.2	95.3	**−29.33**	**−1.56**	**−0.73**
7	3	3	3	6934.5	152.1	16.1	–	3.42	6867.1	6881.7	13.2	0.19	150.1	80.4	–	2.42	**−0.97**
8	3	3	3	7721.8	152.3	56.7	–	−0.45	7609.4	7627.5	24.4	0.32	150.1	85.6	–	**−1.90**	**−1.46**
9	3	3	3	6564.2	152.2	7.0	–	−1.72	6564.2	6592.2	20.8	0.32	150.1	117.6	–	**−1.72**	**0.00**
10	3	3	3	6356.8	152.2	52.3	–	0.39	6318.3	6323.2	2.5	0.04	150.1	68.5	–	**−0.21**	**−0.61**
11	3	5	2	9672.2	153.0	11.7	–	−2.64	9665.6	9797.0	87.2	0.89	150.7	113.8	–	**−2.71**	**−0.07**
12	3	5	2	10837.9	155.8	62.2	–	−1.49	10670.9	10782.8	89.9	0.83	152.2	110.1	–	**−3.00**	**−1.54**
13	3	5	2	9354.7	153.4	21.8	–	−3.60	9441.1	9538.6	78.3	0.82	152.2	125.3	–	**−2.71**	0.92
14	3	5	2	10977.9	153.2	37.8	–	−0.58	10902.8	10974.0	47.7	0.43	150.9	118.8	–	**−1.26**	**−0.68**
15	3	5	2	10300.9	153.2	40.3	–	−9.31	10346.0	10453.1	127.5	1.22	151.0	108.6	–	**−8.92**	0.44
16	3	5	3	9528.2	153.2	24.8	–	−0.26	9591.1	9723.0	101.1	1.04	151.3	133.3	–	0.39	0.66
17	3	5	3	11054.4	154.1	20.9	–	−2.08	11396.8	11480.1	54.8	0.48	152.6	108.4	–	0.95	3.10
18	3	5	3	9635.1	152.9	74.1	–	−2.37	9733.5	9835.1	80.3	0.82	150.8	114.4	–	**−1.37**	1.02
19	3	5	3	6624.4	153.3	86.2	–	−0.85	6624.4	6730.6	114.3	1.70	150.5	103.1	–	**−0.85**	**0.00**
20	3	5	3	9685.9	153.1	43.3	–	0.37	9756.8	9842.4	57.9	0.59	151.8	96.8	–	1.10	0.73
21	5	3	2	10261.2	152.5	32.6	–	−0.69	10020.3	10056.0	45.2	0.45	150.2	93.9	–	**−3.02**	**−2.35**
22	5	3	2	10279.3	152.3	7.7	–	−1.89	9953.5	10067.0	130.6	1.30	150.1	96.0	–	**−5.00**	**−3.17**
23	5	3	2	8376.7	152.3	48.9	–	−0.62	8356.9	8383.6	16.3	0.19	150.3	93.6	–	**−0.85**	**−0.24**
24	5	3	2	7532.9	152.2	59.8	–	0.01	7599.9	7624.9	28.6	0.37	150.1	50.6	–	0.90	0.89
25	5	3	2	8612.9	152.3	25.3	–	−0.56	8593.9	8663.8	60.1	0.69	150.2	111.8	–	**−0.78**	**−0.22**
26	5	3	3	8086.8	152.2	33.8	–	−0.36	8031.6	8105.1	65.7	0.81	150.2	123.3	–	**−1.04**	**−0.68**
27	5	3	3	8055.1	152.2	50.0	−17.19	−1.11	7915.5	8031.0	136.4	1.70	150.1	70.3	**−18.63**	**−2.83**	**−1.73**
28	5	3	3	7896.7	152.2	12.1	–	−0.15	7910.6	7918.4	5.4	0.07	150.2	76.7	–	0.03	0.18
29	5	3	3	9660.3	152.2	32.0	–	0.84	9512.2	9598.7	109.9	1.14	150.2	76.1	–	**−0.71**	**−1.53**
30	5	3	3	7302.1	152.2	4.9	–	−0.92	7311.0	7349.4	56.8	0.77	150.1	41.2	–	**−0.79**	0.12
31	5	5	2	11004.6	154.6	17.1	–	−2.94	10991.3	11069.0	60.5	0.55	151.4	113.2	–	**−3.06**	**−0.12**
32	5	5	2	17665.1	154.6	46.5	–	4.14	16774.7	16914.5	75.1	0.44	153.5	123.1	–	**−1.11**	**−5.04**
33	5	5	2	12101.6	153.5	35.7	–	−1.45	12140.0	12219.6	78.4	0.64	151.3	104.1	–	**−1.14**	0.32
34	5	5	2	10302.8	158.1	93.7	–	0.00	10267.9	10433.0	88.0	0.84	153.4	86.0	–	**−0.34**	**−0.34**
35	5	5	2	12104.5	155.1	31.7	–	−4.50	12175.4	12342.8	154.3	1.25	151.4	120.0	–	**−3.94**	0.59
36	5	5	3	12667.7	156.2	52.6	–	0.32	12669.1	12707.0	48.8	0.38	152.6	123.3	–	0.33	0.01
37	5	5	3	12910.1	153.6	104.8	–	−1.81	12802.9	12950.6	150.4	1.16	152.8	126.1	–	**−2.62**	**−0.83**
38	5	5	3	12270.6	154.6	42.0	–	1.62	12268.7	12381.6	93.6	0.76	152.7	136.9	–	1.60	**−0.02**
39	5	5	3	13148.0	153.3	59.6	–	5.03	12476.0	12720.6	151.5	1.19	151.9	125.7	–	**−0.34**	**−5.11**
40	5	5	3	12313.9	153.0	54.1	–	−0.41	12289.4	12436.1	108.0	0.87	151.0	111.3	–	**−0.61**	**−0.20**
Mean				9494.3	153.2	37.9	−18.36	−0.79	9426.4	9513.4	72.1	0.75	151.0	99.6	−19.22	−1.42	−0.62

In turn, SHVNS clearly outperforms the heuristic approach H, yielding better results for roughly 72% of the instances (86 out of 120). The average $DevToH'$ values are −1.42, −0.72, and −1.18 for the sets with 30, 40, and 50 customers, respectively, thereby improving upon the results previously reported in the literature across all sets of larger instances.

Compared with H-VNS, SHVNS obtains average $DevToHVNS$ values of −0.62, −1.27, and −2.27 for the instance sets with 30, 40, and 50 customers, respectively. The performance gap between the two methods grows with the instance size, indicating that SHVNS is particularly effective for larger instances.

Table 5. Computational results for instances with 40 customers

Inst.	$|\mathcal{P}|$	$|\mathcal{C}|$	$|\mathcal{V}|$	H-VNS					SHVNS								
				Best	*CPU*	*CPU**	*Dev*	*DevToH*	*Best'*	*AvgSol*	σ	*CV*	*CPU*	*CPU**	*Dev'*	*DevToH'*	*DevToHVNS*
41	3	3	2	7670.4	152.4	33.0	–	3.70	7419.9	7547.7	80.8	1.07	150.2	82.1	–	0.32	**−3.27**
42	3	3	2	6996.9	152.2	44.7	–	−0.56	6900.3	6985.4	51.8	0.74	150.1	58.9	–	**−1.93**	**−1.38**
43	3	3	2	7031.5	152.3	22.9	−29.50	−3.33	7009.0	7112.2	95.2	1.34	150.3	107.8	**−29.73**	**−3.64**	**−0.32**
44	3	3	2	7984.0	152.2	52.2	–	−0.56	8031.4	8097.8	61.3	0.76	150.2	86.1	–	0.04	0.59
45	3	3	2	7090.5	152.2	19.3	–	−0.21	7060.2	7078.2	15.5	0.22	150.1	69.8	–	**−0.64**	**−0.43**
46	3	3	3	7725.7	152.2	3.9	–	4.74	7353.7	7432.1	60.6	0.82	150.1	91.0	–	**−0.31**	**−4.82**
47	3	3	3	8564.0	152.2	50.5	–	2.06	8360.6	8500.6	109.2	1.28	150.2	100.9	–	**−0.37**	**−2.38**
48	3	3	3	8552.8	152.2	58.8	–	−0.45	8481.2	8519.1	29.4	0.34	150.1	111.4	–	**−1.28**	**−0.84**
49	3	3	3	7369.4	152.2	37.4	–	0.04	7351.8	7386.5	32.4	0.44	150.3	98.1	–	**−0.20**	**−0.24**
50	3	3	3	7260.3	152.2	24.6	–	0.57	7186.4	7278.8	92.4	1.27	150.1	70.7	–	**−0.46**	**−1.02**
51	3	5	2	11238.5	153.9	64.9	–	−3.84	11350.0	11464.0	80.2	0.70	152.4	109.5	–	**−2.89**	0.99
52	3	5	2	12242.7	155.0	55.5	–	−2.81	12352.1	12441.3	60.2	0.48	151.8	123.5	–	**−1.94**	0.89
53	3	5	2	11673.6	153.5	49.2	–	0.43	11699.9	11713.3	15.2	0.13	151.0	129.5	–	0.65	0.22
54	3	5	2	11621.1	153.5	97.4	–	5.30	11253.2	11483.8	255.5	2.22	151.7	102.4	–	1.96	**−3.17**
55	3	5	2	11579.4	157.1	99.9	–	3.23	11171.9	11224.6	45.6	0.41	152.9	95.7	–	**−0.40**	**−3.52**
56	3	5	3	12712.2	152.7	68.8	–	0.83	12501.8	12584.2	43.9	0.35	151.5	120.8	–	**−0.84**	**−1.65**
57	3	5	3	12882.3	152.6	25.4	–	0.64	12922.3	12973.5	53.4	0.41	150.8	124.9	–	0.96	0.31
58	3	5	3	11037.7	153.2	56.2	–	0.67	10970.8	11093.1	102.0	0.92	151.2	112.2	–	0.05	**−0.61**
59	3	5	3	11483.5	152.9	70.1	–	1.86	11477.9	11563.4	103.5	0.90	152.0	127.7	–	1.81	**−0.05**
60	3	5	3	8847.6	153.4	59.4	–	−2.45	8901.9	9183.6	163.5	1.78	152.6	119.6	–	**−1.85**	0.61
61	5	3	2	10837.9	152.4	51.1	–	1.79	10507.4	10610.0	96.2	0.91	150.1	80.9	–	**−1.32**	**−3.05**
62	5	3	2	14392.1	152.6	16.8	–	−1.91	14431.1	14566.7	100.2	0.69	150.2	82.9	–	**−1.64**	0.27
63	5	3	2	14807.8	152.5	91.2	–	0.48	14216.6	14362.9	134.6	0.94	150.2	116.9	–	**−3.54**	**−3.99**
64	5	3	2	12140.1	152.4	33.4	–	3.67	11673.8	11737.3	52.9	0.45	150.2	87.6	–	**−0.31**	**−3.84**
65	5	3	2	10820.4	152.4	38.3	–	−0.63	10677.7	10756.2	61.5	0.57	150.2	106.5	–	**−1.94**	**−1.32**
66	5	3	3	7137.4	152.2	24.3	–	1.07	7122.4	7170.0	31.2	0.43	150.2	97.0	–	0.86	**−0.21**
67	5	3	3	9185.5	153.3	37.2	–	−2.10	9064.2	9185.9	104.6	1.14	150.1	89.2	–	**−3.39**	**−1.32**
68	5	3	3	11306.0	152.5	43.4	–	−0.19	11447.2	11578.8	78.7	0.68	150.2	107.0	–	1.06	1.25
69	5	3	3	11806.2	152.2	43.7	–	3.15	11498.3	11604.3	80.7	0.70	150.2	97.2	–	0.46	**−2.61**
70	5	3	3	9559.1	152.2	19.7	–	1.44	9306.4	9331.2	18.8	0.20	150.2	95.4	–	**−1.24**	**−2.64**
71	5	5	2	14631.2	154.0	83.0	–	0.08	14420.6	14553.7	99.0	0.68	152.2	110.9	–	**−1.36**	**−1.44**
72	5	5	2	14040.1	156.6	54.8	–	0.91	14057.8	14144.1	68.6	0.48	153.2	117.9	–	1.04	0.13
73	5	5	2	13169.4	154.2	65.8	–	−1.08	13114.4	13188.5	53.8	0.41	151.3	118.3	–	**−1.50**	**−0.42**
74	5	5	2	17520.8	154.3	63.6	–	−1.93	17516.1	17610.4	92.7	0.53	151.9	105.0	–	**−1.96**	**−0.03**
75	5	5	2	17905.2	154.2	59.9	–	4.68	16702.9	16816.5	74.1	0.44	151.8	136.0	–	**−2.35**	**−6.71**
76	5	5	3	13608.0	154.0	41.6	–	−0.06	13704.7	13839.0	127.3	0.92	152.9	128.9	–	0.65	0.71
77	5	5	3	16542.5	153.4	54.7	–	1.48	16288.2	16335.6	29.4	0.18	151.3	108.2	–	**−0.08**	**−1.54**
78	5	5	3	12664.6	153.1	13.5	–	0.15	12694.2	12739.1	51.6	0.41	151.5	93.8	–	0.38	0.23
79	5	5	3	14939.7	152.7	18.1	–	0.07	14759.8	14930.9	154.2	1.03	150.9	119.5	–	**−1.13**	**−1.20**
80	5	5	3	18380.7	153.6	125.4	–	2.56	17847.6	18183.7	222.9	1.23	151.9	106.6	–	**−0.41**	**−2.90**
Mean				11424.0	153.1	49.3	−29.50	0.59	11270.2	11372.7	82.1	0.74	151.0	103.7	−29.73	−0.72	−1.27

The mean coefficient of variation (mean CV), obtained by averaging the relative standard deviation values across the 40 instances in each instance set, indicates a very low dispersion between runs, with mean RSD values below 1%. Therefore, the algorithm shows a high degree of consistency in its performance.

Table 6. Computational results for instances with 50 customers

Inst.	$|\mathcal{P}|$	$|\mathcal{C}|$	$|\mathcal{V}|$	H-VNS					SHVNS								
				Best	CPU	CPU*	Dev	DevToH	Best'	AvgSol	σ	CV	CPU	CPU*	Dev'	DevToH'	DevToHVNS
81	3	3	2	10804.5	152.2	27.7	–	−0.37	10849.8	10883.3	32.2	0.30	150.4	118.7	–	0.05	0.42
82	3	3	2	9617.9	152.2	38.1	–	4.65	9018.5	9179.3	110.2	1.20	150.3	122.7	–	**−1.87**	**−6.23**
83	3	3	2	8944.3	152.2	35.4	–	0.06	8853.1	8938.6	124.5	1.39	150.3	104.9	–	**−0.96**	**−1.02**
84	3	3	2	9869.4	152.2	34.2	–	0.06	9666.7	9773.7	159.9	1.64	150.2	124.5	–	**−2.00**	**−2.05**
85	3	3	2	8884.3	152.3	23.8	–	−0.45	8901.1	9070.4	152.5	1.68	150.3	107.7	–	**−0.26**	0.19
86	3	3	3	8274.9	152.1	42.2	–	−0.10	8255.8	8292.7	26.2	0.32	150.2	131.5	–	**−0.33**	**−0.23**
87	3	3	3	10878.8	152.2	24.7	–	2.37	10703.0	10794.4	58.8	0.54	150.2	108.9	–	0.72	**−1.62**
88	3	3	3	8730.2	152.3	48.2	–	4.38	8378.3	8519.6	129.3	1.52	150.3	104.4	–	0.17	**−4.03**
89	3	3	3	8173.3	152.2	80.1	–	−0.57	8108.1	8246.1	110.8	1.34	150.3	95.1	–	**−1.36**	**−0.80**
90	3	3	3	9846.2	152.2	58.8	–	2.95	9530.7	9640.8	89.8	0.93	150.2	103.2	–	**−0.35**	**−3.21**
91	3	5	2	14163.2	152.9	35.5	–	0.21	14185.7	14303.2	73.1	0.51	151.4	134.4	–	0.37	0.16
92	3	5	2	12960.6	153.9	61.4	–	−1.81	12803.2	13030.1	175.5	1.35	152.7	114.7	–	**−3.00**	**−1.22**
93	3	5	2	11709.0	154.2	72.4	–	−2.71	11789.2	12069.8	141.1	1.17	151.9	123.6	–	**−2.04**	0.69
94	3	5	2	10499.1	159.6	51.8	–	1.56	10619.3	10763.4	202.4	1.88	154.6	127.6	–	2.72	1.14
95	3	5	2	10001.4	154.3	79.9	–	−0.86	10203.1	10301.1	115.4	1.12	152.1	98.1	–	1.14	2.02
96	3	5	3	11939.9	153.6	78.8	–	−0.31	11860.1	11990.4	111.6	0.93	152.3	133.2	–	**−0.97**	**−0.67**
97	3	5	3	15741.4	152.9	27.3	–	−1.90	15847.4	15993.6	77.2	0.48	151.9	119.5	–	**−1.24**	0.67
98	3	5	3	14492.0	152.7	30.3	–	1.59	13997.8	14178.8	180.2	1.27	151.8	101.0	–	**−1.88**	**−3.41**
99	3	5	3	12287.7	153.1	47.8	–	−0.12	12274.2	12525.0	135.0	1.08	152.7	115.6	–	**−0.23**	**−0.11**
100	3	5	3	12480.8	154.1	75.5	–	0.86	12094.5	12457.0	200.7	1.61	153.0	129.3	–	**−2.26**	**−3.09**
101	5	3	2	13247.8	152.3	75.1	–	2.98	12438.4	12538.9	79.9	0.64	150.4	129.7	–	**−3.31**	**−6.11**
102	5	3	2	14134.6	152.1	35.4	–	0.13	13644.7	13680.1	20.5	0.15	150.3	130.6	–	**−3.34**	**−3.47**
103	5	3	2	10775.6	152.2	44.3	–	2.23	10552.5	10612.5	69.6	0.66	150.4	131.4	–	0.11	**−2.07**
104	5	3	2	11589.1	152.2	58.0	–	1.01	11431.3	11577.3	121.8	1.05	150.5	123.0	–	**−0.36**	**−1.36**
105	5	3	2	12763.8	152.3	50.0	–	0.65	12052.7	12119.3	79.3	0.65	150.2	129.0	–	**−4.96**	**−5.57**
106	5	3	3	11157.6	152.2	26.8	–	3.10	10909.0	11063.2	111.6	1.01	150.5	107.0	–	0.80	**−2.23**
107	5	3	3	8862.5	152.2	27.5	–	1.30	8646.4	8728.2	71.7	0.82	150.4	105.0	–	**−1.17**	**−2.44**
108	5	3	3	13567.3	152.2	67.1	–	0.89	13225.0	13273.3	36.1	0.27	150.3	115.3	–	**−1.65**	**−2.52**
109	5	3	3	10858.7	152.2	37.8	–	6.43	10253.8	10401.6	108.4	1.04	150.5	124.8	–	0.50	**−5.57**
110	5	3	3	10273.5	152.2	30.3	–	7.09	9783.1	9827.7	44.2	0.45	150.5	117.0	–	1.98	**−4.77**
111	5	5	2	19898.9	153.1	73.3	–	7.69	18352.1	18515.3	159.6	0.86	151.2	132.9	–	**−0.68**	**−7.77**
112	5	5	2	20123.4	152.7	85.1	–	−2.93	19732.2	19876.7	112.0	0.56	152.3	113.0	–	**−4.81**	**−1.94**
113	5	5	2	14635.6	153.8	33.3	–	4.69	14338.0	14412.2	61.1	0.42	152.3	120.5	–	2.56	**−2.03**
114	5	5	2	21874.2	153.2	55.3	–	−0.26	20343.8	20496.8	170.1	0.83	151.5	142.3	–	**−7.24**	**−7.00**
115	5	5	2	14701.6	152.6	69.3	–	−2.76	14506.1	14662.1	105.1	0.72	150.7	126.3	–	**−4.05**	**−1.33**
116	5	5	3	19414.4	154.1	68.4	–	0.63	19211.3	19536.9	184.2	0.94	152.3	125.7	–	**−0.42**	**−1.05**
117	5	5	3	18736.9	153.8	57.1	–	3.84	17949.3	18162.6	171.8	0.95	153.6	137.1	–	**−0.52**	**−4.20**
118	5	5	3	17420.9	152.7	73.6	–	−2.75	17052.1	17256.0	125.9	0.73	151.2	132.2	–	**−4.81**	**−2.12**
119	5	5	3	20718.7	153.5	44.9	–	0.32	19927.0	20059.2	140.4	0.70	153.1	122.1	–	**−3.51**	**−3.82**
120	5	5	3	16829.9	152.9	35.6	–	2.22	16657.9	16776.3	86.1	0.51	151.4	120.9	–	1.18	**−1.02**
Mean				13047.1	153.0	50.6	–	1.15	12723.7	12863.2	111.6	0.91	151.3	120.1	–	**−1.18**	**−2.27**

5   Conclusions

In this paper, we addressed the Integrated Production and Routing Problem. The problem is particularly challenging, as it combines two complex and interdependent problems, namely a lot-sizing problem and vehicle routing problem. Decisions in one phase, whether production planning or vehicle routing, directly affect the other, making it hard to find high-quality solutions within a reasonable time.

A Skewed Hybrid Variable Neighborhood Search (SHVNS) is proposed, with neighborhood structures designed to modify the routing while the production component is obtained from a MIP model based on precomputed routes.

The proposed SHVNS method outperforms existing approaches in the literature, both in terms of solution quality and computational time. Furthermore, its computational efficiency is maintained even for larger instances, where exact methods become infeasible and heuristic approaches reported in the literature require excessive time. The exact model used in this approach is solved very quickly, and its solving time does not increase with the number of customers, since high-quality routing solutions are driven by the VNS mechanisms. This feature enhances the quality of the strategy for addressing large-scale instances.

Acknowledgments. The first author has been supported by FCT – Fundação para a Ciência e Tecnologia, through national funds from MCTES, and by European Social Fund through NORTE2020 – Programa Operacional Regional Norte, within the research grant SFRH/BD/146217/2019. This work has been supported by FCT – Fundação para a Ciência e Tecnologia within the R&D Unit Project Scope UID/00319/ Centro ALGORITMI (ALGORITMI/UM) https://doi.org/10.54499/UID/00319/2025.

References

1. Darvish, M., Kidd, M.P., Coelho, L.C., Renaud, J.: Integrated production-distribution systems: trends and perspectives. Pesquisa Operacional **41**(s1), e246080 (2021). https://doi.org/10.1590/0101-7438.2021.041s1.00246080
2. Gruson, M., Cordeau, J.F., Jans, R.: Split demand and deliveries in an integrated three-level lot sizing and replenishment problem. Comput. Oper. Res. **161**, 106434 (2024). https://doi.org/10.1016/j.cor.2023.106434
3. Hansen, P., Mladenović, N., Todosijević, R., Hanafi, S.: Variable neighborhood search: basics and variants. EURO J. Comput. Optim. **5**(3), 423–454 (2016). https://doi.org/10.1007/s13675-016-0075-x
4. Leite, M., Pinto, T., Alves, C.: A new improved algorithm for a rich production and routing problem. In: Barata, J., Madani, K., Panetto, H. (eds) Innovative Intelligent Industrial Production and Logistics. IN4PL 2025. Communications in Computer and Information Science, vol. 2826. Springer, Cham (2026). https://doi.org/10.1007/978-3-032-15579-5_14
5. Manousakis, E.G., Kasapidis, G.A., Kiranoudis, C.T., Zachariadis, E.E.: An infeasible space exploring matheuristic for the production routing problem. Eur. J. Oper. Res. **298**, 478–495 (2022). https://doi.org/10.1016/J.EJOR.2021.05.037
6. Miranda, P.L., Cordeau, J.F., Ferreira, D., Jans, R., Morabito, R.: A decomposition heuristic for a rich production routing problem. Comput. Oper. Res. **98**, 211–230 (2018). https://doi.org/10.1016/J.COR.2018.05.004
7. Miranda, P.L., Morabito, R., Ferreira, D.: Mixed integer formulations for a coupled lot-scheduling and vehicle routing problem in furniture settings. Infor. Syst. Oper. Res. (INFOR) **57**, 563–596 (2019). https://doi.org/10.1080/03155986.2019.1575686
8. Russell, R.A.: Mathematical programming heuristics for the production routing problem. Int. J. Prod. Econ. **193**, 40–49 (2017). https://doi.org/10.1016/J.IJPE.2017.06.033

9. Tamssaouet, K., Engebrethsen, E., Dauzère-Pérès, S.: Multi-item dynamic lot sizing with multiple transportation modes and item fragmentation. Int. J. Prod. Econ. **265**, 109001 (2023). https://doi.org/10.1016/j.ijpe.2023.109001
10. Yağmur, E., Kesen, S.E.: Integrated production scheduling and vehicle routing problem with energy efficient strategies: mathematical formulation and metaheuristic algorithms. Exp. Syst. Appl. **237**, 121586 (2024). https://doi.org/10.1016/j.eswa.2023.121586

PyCommend VNS: A Multi-objective Python Library Recommendation Framework

Augusto M. P. de Mendonça[1(✉)], Filipe P. Sousa[2], and Igor M. Coelho[1]

[1] Instituto de Computação, Universidade Federal Fluminense, Niterói, RJ, Brazil
`augustompm@id.uff.br, imcoelho@ic.uff.br`
[2] Instituto de Matemática e Estatística, Universidade do Estado do Rio de Janeiro,
Rio de Janeiro, RJ, Brazil
`filipe.sousa@pos.ime.uerj.br`

Abstract. Python's Package Index hosts over 640,000 packages without standardized categorization, making library discovery difficult. This work formulates library recommendation as multi-objective combinatorial optimization balancing linked usage from project dependencies, semantic similarity from package descriptions, and recommendation set size. We implement Multi-Objective General Variable Neighborhood Search with domain-specific neighborhood structures for addition, removal, and swap operations. The framework processes 10,000 PyPI packages and co-occurrence patterns from 24,000 GitHub repositories. Experimental evaluation with 30 independent runs per algorithm across two context libraries shows MO-GVNS achieves significantly higher hypervolume than NSGA-II (37.4% improvement) and MOEA/D (48.7% improvement), while producing more Pareto-optimal solutions. To our knowledge, VNS has not been applied to library recommendation before.

Keywords: Multi-Objective Optimization · VNS · Library Recommendation

1 Introduction

Python's Package Index (PyPI) hosts over 640,000 packages without standardized categorization. Developers reimplement functionality that exists in established libraries due to awareness gaps, as documented by Xu et al. [14]. Unlike Maven Central with its group-id structure, PyPI lacks organized taxonomies, and Auch et al. [1] found only 75% of Maven libraries have tags.

Consider a developer adopting *fastapi* for web development. Discovering *fastapi* itself may occur through documentation or GitHub, but identifying companion libraries like *pydantic*, *uvicorn*, and *starlette* requires ecosystem knowledge. Ouni et al. [9] formulated library recommendation as multi-objective optimization using NSGA-II, but relied on identifier overlap for semantic similarity without incorporating co-occurrence patterns from dependency files.

S. Cavero et al. (Eds.): ICVNS 2025, LNCS 16256, pp. 95–105, 2026.
https://doi.org/10.1007/978-3-032-19582-1_7

This work applies Variable Neighborhood Search to multi-objective library recommendation. The framework processes 10,000 PyPI packages and dependency patterns from 24,000 GitHub repositories, balancing three objectives: linked usage, semantic similarity, and recommendation set size. Three neighborhood structures guide search: addition, removal, and swap operators. A Pareto archive maintains non-dominated solutions. We follow the MO-GVNS framework of Duarte et al. [5], which separates shaking and local search indices in the canonical GVNS form.

Our contributions are: (1) formulation of library recommendation as multi-objective combinatorial optimization with three competing objectives; (2) adaptation of MO-GVNS with domain-specific neighborhoods for package recommendation; (3) experimental comparison with NSGA-II and MOEA/D showing MO-GVNS achieves higher hypervolume and produces more Pareto-optimal solutions.

Experimental evaluation across two context libraries (*numpy*, *pandas*) with 30 independent runs shows MO-GVNS achieves 37.4% higher hypervolume than NSGA-II and 48.7% higher than MOEA/D. Section 2 reviews related work, Sect. 3 formalizes the problem, Sect. 4 details the algorithm, Sect. 5 presents experimental setup, Sect. 6 analyzes results, and Sect. 7 concludes.

2 Related Work

Xie and Pei [13] developed MAPO to mine API usage patterns from repositories. Thung et al. [11] proposed LibRec combining association rules with collaborative filtering. Ouni et al. [9] formulated library recommendation as multi-objective optimization using NSGA-II, balancing linked usage, semantic similarity, and set size. They relied on identifier overlap rather than semantic embeddings.

Multi-objective evolutionary algorithms address problems with competing objectives through Search-Based Software Engineering, as introduced by Harman and Jones [7]. Deb et al. [3] proposed NSGA-II with fast non-dominated sorting and crowding distance. Zhang and Li [15] introduced MOEA/D, decomposing problems into scalar subproblems using Tchebycheff aggregation. Zhou et al. [16] reviewed decomposition-based optimization advances.

Variable Neighborhood Search exploits different neighborhood structures to escape local optima. Hansen et al. [6] surveyed VNS basics and variants including the canonical General VNS form. Duarte et al. [5] proposed MO-GVNS with algorithms for MO-Shake, MO-VND, and MO-Neighborhood-Change. Dahite et al. [2] introduced MOBI/P for MOVNS. This work follows the MO-GVNS framework of Duarte et al. because it separates shaking (k) and local search (l) indices, matching the canonical GVNS form.

3 Problem Formulation

3.1 Data Collection

The framework uses two data sources: package metadata and project dependencies. Package metadata was collected from the top 10,000 most downloaded PyPI packages. Dependency patterns were extracted from 24,000 GitHub repositories with *requirements.txt* or *pyproject.toml* files.

Package descriptions are encoded using Sentence-BERT [10], producing 384-dimensional embeddings. SBERT extends BERT [4] using siamese networks.

Dependencies parsed from GitHub projects construct a co-occurrence matrix. Entry (i, j) counts projects using both packages i and j.

3.2   Multi-objective Optimization Model

Library recommendation is formulated as multi-objective combinatorial optimization. Given a context library x, the goal is to identify a proper subset $S \subset U$ of related libraries from the universe U of available packages, where $x \notin S$ by definition. The problem balances three competing objectives.

Linked Usage (LU) maximizes co-occurrence patterns from real projects, favoring libraries developers frequently combine with the context library. Semantic Similarity (SS) maximizes topical coherence by measuring alignment between semantic embeddings. Recommended Set Size (RSS) minimizes the number of suggested packages.

These objectives conflict. Maximizing linked usage produces larger sets. Maximizing semantic similarity may select coherent libraries that rarely appear together. Minimizing set size opposes both objectives. The multi-objective formulation produces Pareto-optimal solutions rather than using weighted scoring.

3.3   Mathematical Formulation

Let $U = \{p_1, p_2, \ldots, p_n\}$ denote the universe of n available packages, and let $x \in U$ represent the context library. A solution is a binary vector $\mathbf{s} = (s_1, s_2, \ldots, s_n)$ where $s_i = 1$ if package p_i is included and $s_i = 0$ otherwise, with constraint $s_j = 0$ where $p_j = x$.

Let R be the co-occurrence matrix where R_{ij} is the number of projects containing both p_i and p_j. Let $E_i \in \mathbb{R}^{384}$ denote the SBERT embedding for package p_i. The three objectives are:

$$f_{LU}(\mathbf{s}) = \sum_{i=1}^{n} \sum_{j=1}^{n} s_i \cdot s_j \cdot R_{ij} \cdot \mathbf{1}_{[i \neq j]} \tag{1}$$

$$f_{SS}(\mathbf{s}) = \frac{\sum_{i=1}^{n} s_i \cdot (E_i \cdot E_x)}{\| \sum_{i=1}^{n} s_i \cdot E_i \| \cdot \| E_x \|} \tag{2}$$

$$f_{RSS}(\mathbf{s}) = \sum_{i=1}^{n} s_i \tag{3}$$

The optimization problem is:

$$\begin{aligned}
\text{maximize} \quad & f_{LU}(\mathbf{s}), \quad f_{SS}(\mathbf{s}) \\
\text{minimize} \quad & f_{RSS}(\mathbf{s}) \\
\text{subject to} \quad & \mathbf{s} \in \{0, 1\}^n, \quad s_j = 0 \text{ where } p_j = x
\end{aligned} \tag{4}$$

A solution $\mathbf{s}^*$ is Pareto-optimal if no solution $\mathbf{s}'$ improves at least one objective without worsening another.

4 PyCommend MO-GVNS Framework

4.1 Solution Representation

Solutions are binary vectors $\mathbf{s} = (s_1, s_2, \ldots, s_n)$ where $s_i = 1$ means package p_i is included and $s_i = 0$ means excluded. The context library is never included ($s_j = 0$ where $p_j = x$).

4.2 Multi-Objective General VNS (MO-GVNS)

This work implements MO-GVNS following Duarte et al. [5]. The algorithm maintains an archive $\mathcal{A}$ of non-dominated solutions. MO-GVNS combines three procedures: MO-Shake for diversification, MO-VND for local search, and MO-Neighborhood-Change for index management.

The algorithm uses two independent neighborhood indices following the canonical GVNS form [6]: $k \in \{1, \ldots, k_{max}\}$ controls the shaking intensity for diversification, while $l \in \{1, \ldots, l_{max}\}$ controls the local search depth in MO-VND.

The MO-Improvement procedure checks whether a solution $\mathbf{s}'$ contributes new non-dominated points to archive $\mathcal{A}$. Following Duarte et al., improvement occurs if $\mathbf{s}'$ dominates any solution in $\mathcal{A}$ or is non-dominated with respect to all archive members. Archive maintenance requires $O(|\mathcal{A}| \times m)$ comparisons per candidate solution, where $m = 3$ is the number of objectives. To bound complexity, the archive size is limited to $|\mathcal{A}|_{max} = 100$; when exceeded, crowding distance [3] prunes the most crowded solutions in $O(|\mathcal{A}| \log |\mathcal{A}|)$ time. We use continuous pruning during search rather than post-hoc pruning at the end. Continuous pruning adds overhead per iteration but keeps the archive bounded, avoiding memory growth. Post-hoc pruning would reduce overhead but risk losing good solutions during long runs.

Algorithm 1 presents the MO-GVNS framework adapted for library recommendation.

4.3 Neighborhood Structures

Three neighborhood structures are used:
$\mathbf{N_1}$: **Addition Operator.** The addition operator includes one library not in the current set. Formally, $N_1(S) = \{S \cup \{\ell\} : \ell \in U \setminus (S \cup \{x\})\}$. Adding libraries increases LU and SS but also increases RSS.

Candidate libraries are prioritized by a score combining co-occurrence sum and semantic proximity. Only the top 50 candidates from a pre-computed pool of 200 are evaluated, reducing complexity from $O(|U|)$ to $O(50)$ per neighborhood exploration.

Algorithm 1. MO-GVNS for Library Recommendation

Require: Context library x, package universe U, k_{max}, l_{max}, stopping condition
Ensure: Archive $\mathcal{A}$ of non-dominated solutions
1: $\mathbf{s} \leftarrow$ HybridInitialization(x, U)
2: $\mathcal{A} \leftarrow \{\mathbf{s}\}$
3: **repeat**
4: $k \leftarrow 1$
5: **repeat**
6: $\mathbf{s}' \leftarrow$ MO-Shake$(\mathbf{s}, k, \mathcal{A})$
7: $\mathbf{s}'' \leftarrow$ MO-VND$(\mathbf{s}', l_{max}, \mathcal{A})$
8: $(\mathbf{s}, k) \leftarrow$ MO-NeighborhoodChange$(\mathbf{s}, \mathbf{s}'', k, \mathcal{A})$
9: **until** $k > k_{max}$
10: **until** stopping condition
11: **return** $\mathcal{A}$

N_2: Removal Operator. The removal operator generates neighbors by excluding one library from the current recommendation set. Formally, $N_2(S) = \{S \setminus \{\ell\} : \ell \in S, |S| > 1\}$. This operator reduces set size, directly optimizing the RSS objective.

Libraries are prioritized for removal by computing marginal contribution: $\Delta_\ell = R_{x,\ell} + \alpha \cdot \Sigma_{x,\ell}$ where $\alpha = 0.3$. The library with lowest Δ_ℓ is removed first. Complexity is $O(|S|)$ per removal evaluation.

N_3: Swap Operator. The swap operator simultaneously removes one library and adds another, maintaining constant set size. For each library $\ell_{out} \in S$, candidates ℓ_{in} are ranked by semantic similarity $\Sigma_{\ell_{out}, \ell_{in}}$, selecting replacements that maintain topical coherence. The top 10 swap pairs are evaluated per iteration, yielding complexity $O(|S| \times 10)$.

4.4 MO-VND: Multi-objective Variable Neighborhood Descent

The local search phase implements MO-VND following Duarte et al. [5], enhanced with the MOBI/P strategy from Dahite et al. [2]. MO-VND systematically explores neighborhood structures $N_1^l, N_2^l, \ldots, N_{l_{max}}^l$ for local improvement.

For each neighborhood structure, the search generates neighbors and evaluates them against the archive. Following MOBI/P, neighbors that dominate the current reference become the new reference, while non-dominated neighbors are collected for archive update.

The MO-VND procedure resets to $l = 1$ whenever improvement occurs and advances to $l + 1$ otherwise. Local search terminates when $l > l_{max}$.

4.5 MO-Shake: Shaking Procedure

MO-Shake generates a random solution in the k-th shaking neighborhood $N_k^s(\mathbf{s})$, following Hansen et al. [6]. Shaking intensity increases with k: N_1^s flips few bits, larger k flips more bits.

Following Duarte et al. [5], MO-Shake selects a random solution from $N_k^s(\mathbf{s})$ and updates the archive if non-dominated. After shaking, MO-VND is applied starting from $l = 1$. If the resulting solution improves the archive, search restarts with $k = 1$; otherwise k increments.

5   Experimental Setup

5.1   Experimental Environment

Experiments were conducted on a machine with Intel Core i7-1165G7 processor (2.80 GHz, 4 cores) and 16 GB RAM, running Ubuntu 22.04 LTS. The implementation uses Python 3.10.12 with NumPy 1.24.3, Pandas 2.0.3, SciPy 1.11.1, Scikit-learn 1.3.0, and Sentence-Transformers 2.2.2 for SBERT embeddings.

5.2   Dataset Characteristics

The evaluation uses two data structures from Sect. 3.1. The co-occurrence matrix $R \in \mathbb{R}^{9997 \times 9997}$ has 8.3% non-zero entries (sparse).

The semantic similarity matrix $\Sigma \in \mathbb{R}^{9997 \times 9997}$ has 94% of entries above 0.1 (dense).

Candidate pools are pre-computed: top 200 by co-occurrence and top 200 by semantic similarity, with 30–40% overlap. The threshold of 200 balances coverage against evaluation cost.

5.3   Context Libraries

Two context libraries represent core Python data science packages: *numpy* (numerical computing) and *pandas* (data manipulation). These are fundamental libraries with high connectivity in the co-occurrence matrix, providing diverse recommendation scenarios.

5.4   Algorithm Parameters

Parameters were selected through grid search over 10 preliminary runs per configuration. For MO-GVNS: $k_{max} \in \{2, 3, 5\}$ and $l_{max} \in \{2, 3, 5\}$ were tested. Table 1 shows the results. The combination $k_{max} = 3$, $l_{max} = 3$ achieved highest median hypervolume. Lower values ($k_{max} = 2$) converged faster but to lower quality. Higher values ($k_{max} = 5$) increased runtime without improving results. Archive limit 100 balances diversity against comparison cost. The algorithm terminates after 30 iterations or 10 consecutive iterations without archive improvement.

All algorithms operate under a fixed time budget of 15 s per run. We tested budgets of 10, 15, 30, and 60 s; the 15-s budget provided a good trade-off between solution quality and practical execution time, as longer budgets yielded diminishing returns in hypervolume improvement.

Table 1. Parameter sensitivity for MO-GVNS (10 preliminary runs). Relative HV is normalized to the baseline configuration ($k_{max} = 2$, $l_{max} = 2$).

k_{max}	l_{max}	Relative HV	Time (s)
2	2	1.00	15.2
2	3	1.04	17.8
3	2	1.07	18.4
3	**3**	**1.12**	**21.6**
3	5	1.10	26.3
5	3	1.09	28.1
5	5	1.08	32.7

NSGA-II uses population 50, crossover rate 0.9, mutation rate 0.1. MOEA/D uses population 50, Tchebycheff decomposition, neighborhood size 10, and replacement threshold 3.

All algorithms use hybrid initialization: 40% solutions from co-occurrence pool, 40% from semantic similarity pool, 20% combining both, with 5 initial packages per solution.

5.5   Performance Metrics

Two metrics assess solution quality: hypervolume measures convergence and diversity, spacing measures distribution uniformity. We use the hypervolume algorithm from While et al. [12].

Table 2. Performance comparison across 60 runs per algorithm (30 runs × 2 packages). Bold indicates best value. Values shown as mean ± standard deviation.

	MO-GVNS	NSGA-II	MOEA/D
Hypervolume ↑	**375.89** ±115.07	273.56 ±117.60	252.79 ±54.80
Pareto solutions ↑	**18.8** ±5.1	10.9 ±3.5	13.5 ±0.5
Spacing ↓	588.60 ±332.14	514.17 ±364.40	**475.60** ±72.82
Time (s)	16.1 ±0.7	15.1 ±0.1	15.1 ±0.1

Hypervolume (HV). Hypervolume measures both convergence toward the Pareto front and diversity of solutions. For a solution set A, HV is defined as the volume of objective space dominated by A with respect to a reference point. The reference point is set to $(0, 0, 15)$ after normalizing objectives to $[0, 1]$. Higher hypervolume indicates better performance.

Spacing (Δ). Spacing quantifies the distribution uniformity of solutions along the Pareto front:

$$\Delta = \frac{d_f + d_l + \sum_{i=1}^{|A|-1} |d_i - \bar{d}|}{d_f + d_l + (|A| - 1)\bar{d}} \tag{5}$$

where d_f and d_l are distances to extreme solutions, d_i is the distance between consecutive solutions, and $\bar{d}$ is the mean distance. Lower spacing indicates more uniform distribution.

6   Results and Discussion

6.1   Convergence Analysis

Table 2 presents results across two context libraries with 30 independent runs each (60 total per algorithm). Statistical significance was assessed using Wilcoxon signed-rank tests.

MO-GVNS achieves 37.4% higher hypervolume than NSGA-II (375.89 vs 273.56, $p < 0.001$) and 48.7% higher than MOEA/D (375.89 vs 252.79, $p < 0.001$). The hypervolume difference reflects the elitist archive management in MO-GVNS, which preserves all non-dominated solutions discovered during search.

MO-GVNS produces more Pareto-optimal solutions (18.8) compared to NSGA-II (10.9) and MOEA/D (13.5), indicating better exploration of the objective space. The higher number of solutions contributes to MO-GVNS's higher spacing value (588.60 vs 514.17 and 475.60), as solutions are distributed over a wider region of the Pareto front rather than clustered in a small area. This trade-off between coverage and uniformity is expected when prioritizing exploration over concentration.

Figure 1 shows the hypervolume distribution across 60 runs per algorithm (30 runs for each of the two context libraries). MO-GVNS achieves higher median and mean values than both NSGA-II and MOEA/D. The interquartile range indicates MO-GVNS's consistent performance across different runs, while NSGA-II shows higher variance.

6.2   Recommended Library Sets

Table 3 shows Pareto-optimal recommendations from MO-GVNS. Set sizes range from 3 to 6 libraries. These recommendations match patterns in *requirements.txt* files from the dataset. The framework can generate recommendations for any context library; we show results for the two evaluated contexts plus illustrative examples for other domains.

The Pareto front structure allows developers to select recommendations based on their needs. Smaller sets (3 libraries) provide core companions for minimal projects. Larger sets (5–6 libraries) offer comprehensive toolkits for complex applications.

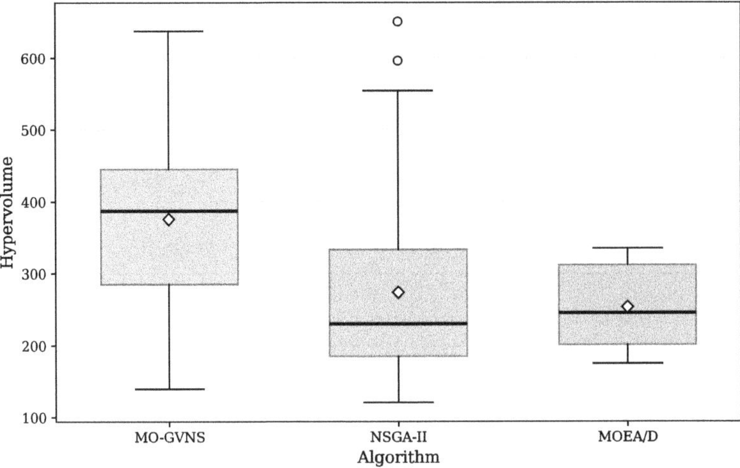

Fig. 1. Hypervolume distribution across 60 runs per algorithm (diamonds indicate means).

6.3 Computational Performance

Under the 15-s time budget, MO-GVNS completes in 16.1 ± 0.7 s (mean $\pm$ std across 60 runs), while NSGA-II and MOEA/D execute in 15.1 ± 0.1 s each. The slight overhead in MO-GVNS comes from the VND local search phase.

Source code is available at GitHub[1].

7 Conclusions

This work presented PyCommend, a library recommendation framework using Multi-Objective General Variable Neighborhood Search. The algorithm balances three objectives—linked usage, semantic similarity, and recommendation set size—through domain-specific neighborhoods for package addition, removal, and swap.

Experimental evaluation with 30 independent runs per algorithm across two context libraries demonstrates that MO-GVNS significantly outperforms NSGA-II and MOEA/D in hypervolume, with improvements of 37.4% and 48.7% respectively ($p < 0.001$). MO-GVNS also discovers more Pareto-optimal solutions, indicating better exploration of the trade-off space. Generated recommendations align with co-occurrence patterns observed in real project dependencies.

---

[1] https://github.com/augustompm/pycommend-vns.

Table 3. Example Pareto-optimal recommendations from MO-GVNS

Context	Recommended Set	Size
numpy	*scipy, matplotlib, pandas*	3
numpy	*scipy, matplotlib, pandas, scikit-learn*	4
numpy	*scipy, matplotlib, pandas, scikit-learn, pillow, h5py*	6
pandas	*numpy, matplotlib, seaborn*	3
pandas	*numpy, matplotlib, seaborn, scikit-learn, openpyxl*	5
pandas	*numpy, matplotlib, seaborn, scikit-learn, openpyxl, sqlalchemy*	6
fastapi	*pydantic, uvicorn, starlette*	3
fastapi	*pydantic, uvicorn, starlette, httpx, pytest, sqlalchemy*	6
scikit-learn	*pandas, matplotlib, seaborn, xgboost*	4
scikit-learn	*pandas, matplotlib, seaborn, xgboost, optuna*	5

7.1 Limitations and Future Work

The current framework covers 10,000 PyPI packages and does not model version compatibility, security vulnerabilities, or license constraints. Semantic similarity depends on PyPI descriptions, which vary in quality across packages. The experimental evaluation used two context libraries; broader evaluation with additional domains would strengthen generalization claims.

Future work includes integration with vulnerability databases for security-aware recommendations and extension to other ecosystems (npm, Maven, Cargo). Developer evaluation studies through A/B testing could measure real-world impact on productivity. Automatic parameter tuning with tools like irace could adapt algorithm configuration to different problem instances. Recent work by Liu et al. [8] suggests potential integration of large language models for adaptive search guidance in multi-objective optimization, which could improve neighborhood selection strategies.

Acknowledgments. This work was supported by CAPES - Finance Code 001, CNPq (grant 312165/2025-0), and FAPERJ (grant JCNE 2021 E-26/201.451/2022).

References

1. Auch, M., Balluf, M., Mandl, P., Wolf, C.: Towards an automated classification of software libraries. SN Comput. Sci. **5**(339) (2024). https://doi.org/10.1007/s42979-024-02654-2
2. Dahite, L., Kadrani, A., Benmansour, R., Guibadj, R.N., Fonlupt, C.: Multi-objective model and variable neighborhood search algorithms for the joint maintenance scheduling and workforce routing problem. Mathematics **10**(11), 1807 (2022). https://doi.org/10.3390/math10111807

3. Deb, K., Pratap, A., Agarwal, S., Meyarivan, T.: A fast and elitist multiobjective genetic algorithm: NSGA-II. IEEE Trans. Evol. Comput. **6**(2), 182–197 (2002). https://doi.org/10.1109/4235.996017
4. Devlin, J., Chang, M.W., Lee, K., Toutanova, K.: BERT: pre-training of deep bidirectional transformers for language understanding. In: Proceedings of the 2019 Conference of the North American Chapter of the Association for Computational Linguistics: Human Language Technologies, Volume 1 (Long and Short Papers), pp. 4171–4186. Association for Computational Linguistics (2019). https://doi.org/10.18653/v1/N19-1423
5. Duarte, A., Pantrigo, J.J., Pardo, E.G., Mladenović, N.: Multi-objective variable neighborhood search: an application to combinatorial optimization problems. J. Global Optim. **63**(3), 515–536 (2015). https://doi.org/10.1007/s10898-014-0213-z
6. Hansen, P., Mladenović, N., Todosijević, R., Hanafi, S.: Variable neighborhood search: basics and variants. EURO J. Comput. Optim. **5**(3), 423–454 (2017). https://doi.org/10.1007/s13675-016-0075-x
7. Harman, M., Jones, B.F.: Search-based software engineering. Inf. Softw. Technol. **43**(14), 833–839 (2001). https://doi.org/10.1016/S0950-5849(01)00189-6
8. Liu, W., Chen, L., Tang, Z.: Large language model aided multi-objective evolutionary algorithm: a low-cost adaptive approach (2024)
9. Ouni, A., Kessentini, M., Inoue, K., Cinneide, M.O.: Search-based software library recommendation using multi-objective optimization. Inf. Softw. Technol. **83**, 55–75 (2017). https://doi.org/10.1016/j.infsof.2016.11.007
10. Reimers, N., Gurevych, I.: Sentence-BERT: sentence embeddings using Siamese BERT-networks. In: Proceedings of the 2019 Conference on Empirical Methods in Natural Language Processing and the 9th International Joint Conference on Natural Language Processing (EMNLP-IJCNLP), pp. 3982–3992. Association for Computational Linguistics (2019). https://doi.org/10.18653/v1/D19-1410
11. Thung, F., Lo, D., Lawall, J.: Automated library recommendation. In: 2013 20th Working Conference on Reverse Engineering (WCRE), pp. 182–191. IEEE (2013). https://doi.org/10.1109/WCRE.2013.6671293
12. While, L., Hingston, P., Barone, L., Huband, S.: A faster algorithm for calculating hypervolume. IEEE Trans. Evol. Comput. **10**(1), 29–38 (2006). https://doi.org/10.1109/TEVC.2005.851275
13. Xie, T., Pei, J.: MAPO: mining API usages from open source repositories. In: 2006 International Workshop on Mining Software Repositories, MSR '06, pp. 54–57. ACM (2006). https://doi.org/10.1145/1137983.1137997
14. Xu, B., Ye, D., Xing, Z., Xia, X., Chen, G., Li, S.: Why reinventing the wheels? An empirical study on library reuse and re-implementation. Empir. Softw. Eng. **25**, 755–789 (2020). https://doi.org/10.1007/s10664-019-09771-0
15. Zhang, Q., Li, H.: MOEA/D: a multiobjective evolutionary algorithm based on decomposition. IEEE Trans. Evol. Comput. **11**(6), 712–731 (2007). https://doi.org/10.1109/TEVC.2007.892759
16. Zhou, A., Qu, B., Li, H., Zhang, Q.: A survey of decomposition-based evolutionary multi-objective optimization: part II - a data science perspective (2024)

Multi-objective VNS Tourism Planning: Optimizing Weekend Itineraries in Montreal

Filipe P. Sousa[1]([✉]) [iD], Augusto M. P. de Mendonça[2] [iD], and Igor M. Coelho[2] [iD]

[1] COMPMAT- Instituto de Matemática e Estatística, Universidade do Estado do Rio de Janeiro, Rio de Janeiro, RJ, Brazil
filipe.sousa@pos.ime.uerj.br
[2] Instituto de Computação, Universidade Federal Fluminense, Niterói, RJ, Brazil
augustompm@id.uff.br, imcoelho@ic.uff.br

Abstract. This paper addresses the Tourist Trip Design Problem (TTDP) in an urban context, focusing on weekend itineraries in Montreal, Canada. The problem is modeled as a multi-objective combinatorial optimization task that incorporates real-world constraints such as attraction opening hours, transportation modes, budget, and user preferences. A Multi-objective Variable Neighborhood Search (MOVNS) metaheuristic is applied, combining seven neighborhood operators with an adaptive shake strategy to explore diverse regions of the Pareto front. As a reference method, we consider the Non-dominated Sorting Genetic Algorithm II (NSGA-II) under the same computational budget. Both algorithms are evaluated on realistic instances built from GTFS, OpenStreetMap, and online tourist platforms.

Experimental results show that MOVNS achieves higher Hypervolume and lower Inverted Generational Distance than NSGA-II, while maintaining competitive Spread and additive epsilon-indicator values. Taken together, these indicators suggest that MOVNS provides a broader and more informative approximation of the Pareto front for the same time budget. A qualitative analysis also shows that the algorithm can generate customized itineraries for different traveler profiles (economic vs. premium). Overall, the results indicate that MOVNS is suitable for real-world tour planning scenarios and can serve as a core component of intelligent recommendation systems in urban tourism.

Keywords: Tourist Trip Design Problem · Multi-objective Optimization · Variable Neighborhood Search · Urban Mobility · Itinerary Planning · Metaheuristics

1 Introduction

The planning of personalized tourist itineraries represents a relevant challenge in operations research, as it integrates combinatorial optimization, urban mobility, and user preferences. With the growth of urban tourism and the demand for

S. Cavero et al. (Eds.): ICVNS 2025, LNCS 16256, pp. 106–119, 2026.
https://doi.org/10.1007/978-3-032-19582-1_8

customized experiences, it has become essential to develop models capable of generating efficient itineraries while respecting time, budget, and transportation constraints.

The *Tourist Trip Design Problem* (TTDP) models this scenario by extending the *Orienteering Problem* to include time windows, multimodal transportation, and individual preferences [2,12]. The literature highlights the use of meta-heuristics as effective strategies to address its complex and highly constrained nature. Studies such as those by Schilde et al. [8] and Son et al. [9] have explored problem variants through hybrid algorithms such as ACO, VNS, and GA. Approaches such as GLNPSO [14] and ALNS [3] have also achieved competitive performance, particularly in formulations that emphasize sustainability, cost, and user satisfaction.

Environmental concerns have also gained relevance: green models for urban routing, such as those proposed by Divsalar et al. [1], aim to reduce emissions and promote energy efficiency.

In this context, this study proposes the application of the *Multi-objective Variable Neighborhood Search* (MOVNS) metaheuristic to the planning of two-day tourist itineraries in Montreal, Canada. The approach simultaneously optimizes the four objectives F_1–F_4: F_1 (number of attractions visited), F_2 (overall quality score), F_3 (total route time), and F_4 (total trip cost). Real data on hotels and attractions are used, including opening hours, reviews, and travel information derived from GTFS and OpenStreetMap.

In this study, we (i) apply MOVNS to an urban instance with multimodal transportation (metro, bus, car, and walking) and (ii) compare it with NSGA-II under the same time budget, focusing on how each algorithm covers the Pareto frontier and handles solution diversity.

2   Literature Review

2.1   Classical Models and Single-Objective Formulations

The *Tourist Trip Design Problem* (TTDP) is a generalization of the *Orienteering Problem* (OP), in which the goal is to select a subset of attractions that maximizes attractiveness within a limited time or cost budget [12]. Early TTDP formulations adopted single-objective models, usually focused on maximizing aggregate satisfaction under strict time and budget constraints [2]. Integer linear programming proved effective only for small instances due to the high combinatorial complexity [7].

As the problem matured, TTDP models began to incorporate additional real-world aspects such as time windows, clustering of attractions, and multiple transportation modes [11,14]. These extensions improved realism and practical relevance but also increased the size and complexity of the search space. The limitations of single-criterion models became evident when balancing conflicting objectives such as cost, time, and quality, which motivated the transition to multi-objective formulations more aligned with user preferences and diverse tourism contexts [1,5].

2.2 Metaheuristics in the TTDP

The complexity of realistic TTDP instances motivated the use of metaheuristics to obtain high-quality solutions within reasonable computation times. Among the earliest approaches, the *Greedy Randomized Adaptive Search Procedure* (GRASP) was applied by Santos-Peñate et al. [7], while *Variable Neighborhood Search* (VNS), including bi-objective variants, was explored in works such as [8,9]. VNS stands out for its ability to escape local optima and promote diversity in the search process.

Hybrid strategies were also proposed, such as the combination of *Ant Colony Optimization* (ACO) and VNS [8], and the use of *Adaptive Large Neighborhood Search* (ALNS) in sustainable tourism settings [3]. Wisittipanich and Boonya [14] adapted *Particle Swarm Optimization* (PSO) to explore urban search spaces efficiently, while Zeine and Zeina [16] investigated the *Backtracking Search Algorithm* (BSA) in related routing contexts.

Comparative analyses of multi-objective strategies [15] emphasize the joint evaluation of convergence and diversity, reinforcing the role of indicators such as Hypervolume and IGD. In green variants, Divsalar et al. [1] proposed models that explicitly minimize emissions, and Raoui et al. [6] discussed the ethical implications of using metaheuristics in decision-making processes.

2.3 Multi-objective Formulations

The natural evolution of the TTDP led to multi-objective formulations, better suited to the complexity of real tourism scenarios. Schilde et al. [8] studied the *Bi-objective Orienteering Problem*, balancing attractiveness and travel time. From this perspective, algorithms such as the *Non-dominated Sorting Genetic Algorithm II* (NSGA-II) gained prominence due to their robustness and capacity to generate diverse Pareto fronts [4], and have been widely used in multi-objective TTDP variants.

The *Multi-objective Variable Neighborhood Search* (MOVNS), a natural extension of VNS to multi-objective problems, was investigated by Son et al. [9] as a promising alternative to explore multiple neighborhoods while preserving diversity. Sustainable and socially aware formulations, such as those proposed by Pitakaso et al. [5], incorporate environmental and social criteria into itinerary design. Other implementations, including *Global Local and Near-Neighbour Particle Swarm Optimization* (GLNPSO) [14], have shown good performance in real urban environments.

Structured comparisons between algorithms, such as those by Zavala et al. [15], highlight the impact of methodological choices on convergence and diversity. Surveys such as Liu et al. [4] consolidate the TTDP as a challenging domain for multi-objective metaheuristics, due to its rich combination of operational constraints, user preferences, and sustainability concerns.

In summary, the evolution of the TTDP reflects a growing demand for personalized, sustainable, and adaptive solutions, in which multi-objective metaheuristics play a central role.

3   Methodology

3.1   Problem Formulation

The *Tourist Trip Design Problem* (TTDP) aims to generate tourist itineraries that maximize user satisfaction when visiting a subset of attractions while respecting logistical and operational constraints. It is a combinatorial, multi-objective problem whose complexity grows rapidly with the number of points of interest [2, 12].

In this study, the TTDP is modeled as a directed graph $G = (V, A)$, where $V = \{v_0, ..., v_n, v_{n+1}\}$ represents the vertices (attractions and hotels) and A the arcs (transfers). Each vertex has an attractiveness score a_i, a service time s_i, and a time window $[e_i, l_i]$ [3, 10].

We use the following decision variables. For each attraction $i \in V$, a binary variable $x_i \in \{0, 1\}$ indicates whether attraction i is visited ($x_i = 1$) or not ($x_i = 0$). For each feasible transfer $(i, j) \in A$, a binary variable $y_{ij} \in \{0, 1\}$ is equal to 1 if the itinerary travels directly from location i to location j. In addition, a continuous variable T_i represents the starting service time at attraction i and waiting times caused by early arrival are incorporated into the total route time.

In this study we optimize four objectives simultaneously: maximize the number of visited attractions, $F_1(x) = \sum_i x_i$; maximize total attractiveness (quality score), $F_2(x) = \sum_i a_i x_i$; minimize total route time, $F_3(x, y) = \sum_{(i,j)} t_{ij} y_{ij}$; and minimize total monetary cost, $F_4(x, y) = \sum_{(i,j)} c_{ij} y_{ij}$.

Waiting times caused by early arrival are included in the total route time F_3.

For clarity, we refer to these objectives as F_1 (number of attractions), F_2 (quality score), F_3 (route time) and F_4 (cost), and use the same notation in the descriptive tables of representative itineraries.

The constraints include: (i) a limit on the total available time; (ii) respect for visitation time windows; (iii) elimination of subtours; and (iv) modal compatibility between transfers, considering metro, bus, car, and walking [5].

Given its multi-objective nature, the goal is to generate a set of non-dominated solutions that approximate the Pareto frontier, enabling the end user to select itineraries with different trade-offs among attractiveness, time, cost, and comfort [4, 6].

3.2   Experimental Setup

This study was conducted in the city of Montreal, Canada, chosen for its tourism diversity and the availability of open urban data. We consider two-day tourist itineraries with 12-h daily windows, starting and ending at hotels located in the downtown area (Ville-Marie), ensuring distinct yet comparable routes [12].

Instances were built from real data:

– **Attractions**: extracted via Google Places, TripAdvisor and OpenStreetMap APIs, with attractiveness computed from average ratings, number of reviews and thematic category, following [3, 10].

- **Urban transportation**: GTFS data from STM and walking times from OpenRouteService, accounting for multimodal transfers, penalties and real schedules [5].
- **Hotels**: fixed departure/return points each day, selected to offer good accessibility to the transport network.

The final Montreal instance comprises $|V| = 150$ locations, including $N_{att} = 50$ tourist attractions and $N_{hot} = 100$ candidate hotels, and $|A| = 37\,491$ feasible directed transfer arcs derived from GTFS and street network data. The planning horizon covers two consecutive days with a 12-h window per day.

The computational experiments were implemented in Python 3.11 using `NumPy`, `NetworkX`, `GTFSKit` and `OSMNX` to integrate spatial data, public transport information and the optimization algorithms. All runs were executed on an Ubuntu 22.04 (64-bit) system with an Intel Core i7-11800H processor (8 physical cores) and 32 GB of RAM.

Both algorithms were executed under the same wall-clock time budget of 240 s per run, controlled by a common `--max-time` parameter, and each configuration was run 30 times with different random seeds for statistical comparison. For MOVNS, we set `--solutions = 4`, `--archive-max = 60` and kept an idle-iteration safeguard (`--no-improvement = 2`), although in practice the time budget is the dominant stopping criterion. NSGA-II was run with a population of 40 individuals and 20 generations, with crossover probability 0.6 and mutation probability 0.5. In both cases the algorithms stop when either the time budget is reached or their internal limits are met. Random seeds ensure reproducibility, and the data are organized for future public release, following good practices of open and reproducible science [4,6,8].

3.3 Solution Representation

Each solution is represented by two ordered vectors $D_1 = (v_{1,1}, v_{1,2}, \dots)$ and $D_2 = (v_{2,1}, v_{2,2}, \dots)$, where each $v_{d,i}$ corresponds to a distinct attraction, reflecting the chronological sequence of visits on day 1 and day 2 of planning, respectively. This structure directly encodes the order of the tourist points to be visited on each day.

Transfers between consecutive pairs $(v_{d,i}, v_{d,i+1})$ are computed based on real multimodal transportation data from Montreal, obtained from precomputed travel-time and cost matrices built using GTFS data from STM and the street network. The modal choice (car, metro, bus, or walking) is explicitly encoded in the solution and used to index these matrices, prioritizing minimum travel time and adequate geographic coverage [5,10].

Temporal feasibility is ensured through auxiliary structures that store estimated arrival, service, and departure times for each point, respecting opening hours, minimum visit durations, and the daily limit of 12 h. In dynamic operations (such as mutations), linked lists are used to facilitate insertions and removals without compromising solution consistency [6,8].

Solution evaluation considers four objectives, computed from real data and individual preferences, as described earlier. Comparisons between individuals are performed via Pareto dominance, with non-dominated solutions stored in an external archive updated at each iteration [2,3,12].

This representation ensures flexibility and adherence to the urban context, enabling the construction of realistic and personalized itineraries for different tourist profiles.

3.4   MOVNS

The proposed algorithm is based on the *Multi-objective Variable Neighborhood Search* (MOVNS), adapted to the *Tourist Trip Design Problem* (TTDP). MOVNS explores several complementary neighborhoods under an external elitist archive that stores the current approximation of the Pareto front, which is well suited to the highly constrained and non-convex search space of urban routing problems [3,4,6].

Seven neighborhood operators are employed:

- N1: swap of two attractions within the same day;
- N2: move of an attraction between days;
- N3: insertion or removal of an attraction;
- N4: replacement of an attraction by another not yet visited;
- N5: 2-opt reversal of a route segment;
- N6: change of departure/arrival hotel, followed by route re-evaluation;
- N7: change of transport mode on a given segment.

The initial set of solutions P is small and controlled by a parameter (`--solutions`). A subset of P is built by greedy constructive heuristics that insert attractions with high $\frac{a_k}{\Delta t_k}$ values and reasonable geographic proximity, while the remaining solutions are generated at random, provided that time-window and budget constraints are respected. The non-dominated solutions of P form the initial external archive A. During the search, every new non-dominated solution is inserted into A, dominated solutions are removed, and if $|A| > 60$ the solution with the smallest marginal hypervolume contribution is discarded.

Diversification is driven by a *shake* operator that performs a random walk of length k in the neighborhood space. Starting from a solution $S \in A$, a sequence of k moves is sampled from $N_1, \ldots, N_7$ with uniform probability and applied only if feasibility with respect to time windows and budget is preserved. The perturbed solution S' is then evaluated and possibly inserted into the archive.

The parameter $k \in \{1, \ldots, 5\}$ controls the perturbation strength. Whenever S' improves the archive (by dominating some solution or entering as a new non-dominated point), k is reset to 1, emphasizing intensification around promising regions. Otherwise, k is incremented up to $k_{\max} = 5$, increasing diversification. Preliminary experiments with larger values of k produced overly disruptive moves and higher runtimes without consistent hypervolume gains, which motivated the choice of $k_{\max} = 5$ as a compromise.

Solution evaluation simultaneously considers four objectives: (i) maximization of the number of attractions visited (F_1); (ii) maximization of the overall quality score (F_2); (iii) minimization of total route time (F_3); and (iv) minimization of total trip cost (F_4). MOVNS is primarily stopped by a wall-clock limit of 240 s per run (parameter `--max-time`); secondary safeguards based on the number of consecutive iterations without improvement in A are also monitored.

The general structure of the algorithm is summarized in Algorithm 1.

Algorithm 1. MOVNS for the TTDP

1: Generate initial feasible population P (size controlled by `--solutions`)
2: Initialize external archive $A \leftarrow$ non-dominated solutions of P
3: $k \leftarrow 1$
4: **while** stopping criterion not satisfied **do**
5:     Select solution $S \in A$ via round-robin
6:     Select neighborhood N_i randomly
7:     Apply *shake* with strength k on S using N_i, obtaining S'
8:     **if** S' is feasible **then**
9:         **if** S' dominates some solution in A or is non-dominated w.r.t. A **then**
10:             $A \leftarrow A \cup \{S'\}$, removing dominated solutions
11:             $k \leftarrow 1$
12:         **else**
13:             $k \leftarrow \min(k + 1, 5)$
14:         **end if**
15:         **if** $|A| > 60$ **then**
16:             Remove solution with lowest hypervolume contribution
17:         **end if**
18:     **end if**
19: **end while**
20: **return** archive A as Pareto-front approximation

3.5   Comparison Algorithm

As a comparison strategy, we adopted the Non-dominated Sorting Genetic Algorithm II (NSGA-II), one of the most established approaches for multi-objective optimization. Since its original proposal by Deb et al. (2002), NSGA-II has been widely used as a baseline due to its robustness and its ability to generate diverse Pareto fronts in complex domains, including routing and urban planning problems [4,15]. Its operation combines non-dominated sorting, crowding distance and elitist selection, which together promote convergence towards promising regions of the search space while preserving diversity.

In this work, NSGA-II was implemented directly in Python, in the `nsga2` module of the project, and adapted to the same solution encoding used by MOVNS, ensuring direct comparability. The initial population is built by guided randomized construction heuristics that produce feasible itineraries respecting time windows and budget constraints. Crossover is performed using two-point ordered crossover (OX) with probability 0.60, while mutation is applied with probability 0.50 using problem-specific moves, including hotel swap, addition, removal or replacement of attractions, and changes in day or transport mode. Selection follows a binary tournament scheme, with ties broken by crowding distance, and elitism is guaranteed by retaining the set of non-dominated solutions at each generation.

Parameter values for population size and operator rates follow recommendations from the NSGA-II literature and recent parameterization studies that emphasize the use of sufficiently large populations and standard crossover/mutation settings for robust performance [13], and were then refined by preliminary tuning on the Montreal instance under the same time budget as MOVNS. The final configuration uses a population of 40 individuals, 20 generations, crossover probability 0.60 and mutation probability 0.50. NSGA-II is executed under the same 240-second wall-clock limit as MOVNS, and terminates when either this time budget is reached or the maximum number of generations is completed.

This setup allows the relative performance of MOVNS to be assessed against a widely accepted baseline under identical computational conditions, in terms of Pareto-front quality, solution diversity and stability across multiple runs [6,9].

3.6   Evaluation Metrics

The quality of the solution sets produced by MOVNS and NSGA-II is assessed using four standard indicators that jointly capture convergence, diversity, and worst-case proximity to a reference frontier: Hypervolume (HV), Inverted Generational Distance (IGD), Spread, and the additive epsilon indicator.

Hypervolume (HV): Measures the volume of the objective space dominated by a set of non-dominated solutions with respect to a reference point. Higher values indicate better overall quality, as HV reflects both convergence and diversity of the approximated Pareto frontier [4,15].

Inverted Generational Distance (IGD): Computes the average distance from each point of a true (or reference) Pareto frontier to its nearest neighbor in the approximated set. Lower IGD values correspond to better convergence; here, the reference set is obtained from the union of non-dominated solutions produced by all algorithms across all runs [15].

Spread: Evaluates how well the solutions are distributed along the frontier, penalizing gaps and clusters. Lower Spread values indicate a more uniform coverage of the trade-off surface and, consequently, better diversity [4,6].

Additive Epsilon Indicator: Quantifies how much an approximation set must be translated in objective space to weakly dominate a reference set. Lower values mean that the obtained frontier is closer to the reference in a worst-case sense; this indicator is widely used in indicator-based multi-objective search and evolutionary algorithms [17,18] and is computed here with respect to the same reference set used for IGD, serving both for comparative analysis and convergence monitoring.

Each algorithm was executed 30 times with different initialization seeds. For every metric, we report mean and standard deviation, allowing us to evaluate not only central performance but also the stability and variability of the results across runs.

4   Results and Discussion

This section presents the experimental results obtained by MOVNS and NSGA-II for the Tourist Trip Design Problem (TTDP) with four conflicting objectives. The analysis relies on standard multi-objective metrics: Hypervolume (HV), Inverted Generational Distance (IGD), Spread, and the additive epsilon indicator, jointly evaluating solution quality, convergence, diversity and worst-case proximity to a reference Pareto front.

Each algorithm was executed in 30 independent runs with different random seeds. For each metric, we report mean and standard deviation. These values are shown in Table 1, and Fig. 1 provides a visual comparison with error bars.

4.1   Quantitative Evaluation of Solutions

Table 1. Average results (mean ± standard deviation) of the multi-objective metrics over 30 runs.

Algorithm	Hypervolume	IGD	Spread	ϵ-indicator
MOVNS	0.089 ± 0.010	80.0 ± 19.0	1.12 ± 0.04	0.41 ± 0.08
NSGA-II	0.073 ± 0.012	152.0 ± 42.0	1.03 ± 0.05	0.04 ± 0.04

The results reveal a clear trade-off between the two algorithms. MOVNS attains higher Hypervolume and lower IGD, indicating broader coverage of the dominated region and closer convergence to the reference front. NSGA-II, in turn, achieves better Spread and epsilon values, suggesting a more compact and more uniformly spaced front that stays closer to the reference set in a worst-case sense. Overall, MOVNS offers a richer approximation of the Pareto frontier, while NSGA-II concentrates the search in a narrower region with very small worst-case deviations.

Figure 1 reinforces these patterns. The bar chart with error bars presents, for each metric, the mean values, their standard deviations and the average size of the Pareto set. The panels for Hypervolume and IGD clearly favor MOVNS, whereas Spread and epsilon favor NSGA-II. The Pareto panel also shows that MOVNS systematically generates larger sets of non-dominated solutions across runs.

Because MOVNS consistently attains higher Hypervolume and lower IGD, it is a better option when the goal is to cover a broad set of high-quality trade-offs. At the same time, they show that NSGA-II remains a competitive baseline when a more compact set of solutions with very small worst-case deviations is sufficient. The results are consistent with recent literature emphasizing the potential of variable-neighborhood-based metaheuristics for multi-objective combinatorial problems [4–6].

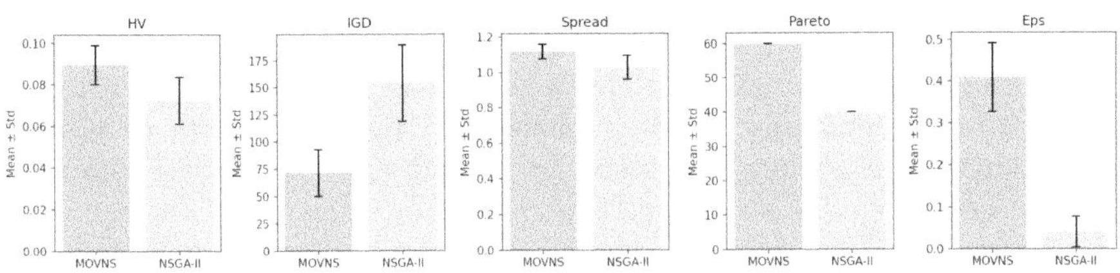

Fig. 1. Comparison between MOVNS and NSGA-II in performance metrics: Hypervolume, IGD, Spread, number of Pareto solutions, and epsilon-indicator (mean ± standard deviation over 30 runs).

4.2   Qualitative Analysis of Itineraries

In addition to the quantitative indicators, a qualitative analysis helps to interpret the practical trade-offs among the objectives. To this end, two representative solutions from the Pareto frontier obtained by MOVNS were selected, illustrating distinct weekend profiles in Montreal.

Table 2 compares an **economic** itinerary (run-240 s-movns-2) and a **premium** itinerary focused on maximizing the number of visits and overall quality (run-240 s-movns-8). The economic solution stays at *Auberge L'Apero* and visits only free attractions, using walking as the exclusive transport mode. The premium solution is based at the *Alternative Hostel of Old Montreal*, combines walking and car segments, and includes paid attractions such as *Oasis Immersion*, which leads to more visits and higher aggregated quality.

Table 2. Comparison between the **Economic** (run-240 s-movns-2) and **Premium** (run-240 s-movns-8) itineraries.

Characteristic	Economic	Premium
Hotel	Auberge L'Apero	Alternative Hostel of Old Montreal
Hotel cost	$25.25	$35.75
Total trip cost	$25.25	$154.34
Number of attractions visited (F_1)	13	14
Quality score (F_2)	66.2	70.3
Total time (minutes)	1332.2	1268.8

The economic itinerary minimizes monetary cost by combining a low-cost hostel with a set of free attractions and walking-only transfers. This choice increases total travel time slightly but still yields a solid quality score, supported by highly rated points of interest. The premium itinerary, in turn, accepts a higher budget to include paid experiences and faster transport segments, allowing a larger number of visits with higher aggregated quality in a slightly shorter total time.

Together, these itineraries illustrate how the four objectives can be balanced according to different user profiles, reinforcing MOVNS's ability to generate diverse and customizable alternatives. To support practical use, an interactive interface was developed for visualizing routes and attractions. Figure 2 shows the main screen, where users can explore itineraries and adjust preferences.

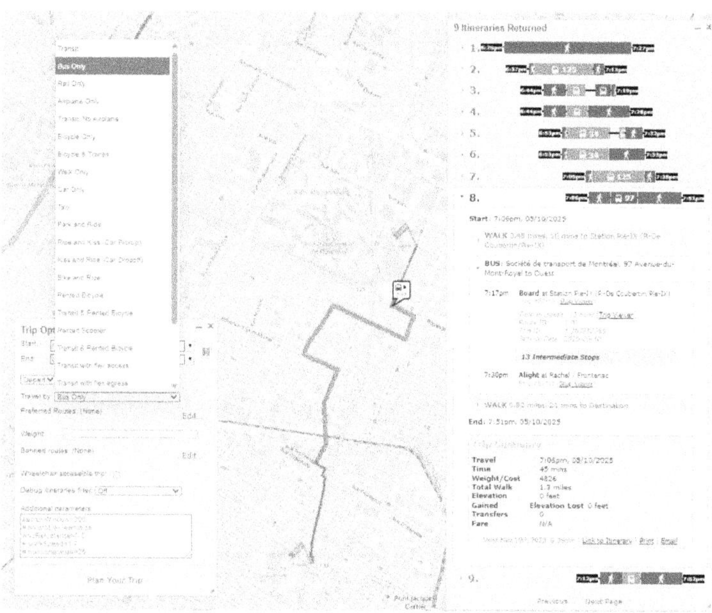

Fig. 2. Visualization of the developed interface for interactive exploration of personalized itineraries generated by MOVNS.

4.3   Algorithmic Performance and Robustness

Beyond solution quality metrics, the computational performance analysis of the algorithms is essential to assess their practical applicability in real-world tourism planning scenarios. In this study, MOVNS and NSGA-II were executed under the same wall-clock time budget of 240 s per run, enforced by the `--max-time` parameter. Table 3 presents a direct comparison between the algorithms across five relevant indicators: average number of non-dominated solutions generated, average Hypervolume, Spread metric, epsilon-indicator, and execution time.

The results indicate that MOVNS achieves superior performance in terms of solution quality and robustness of the obtained Pareto frontier. Although both

Table 3. Performance comparison between MOVNS and NSGA-II (240 s, 30 runs).

Indicator	NSGA-II	MOVNS
Number of Pareto solutions	60	60
Hypervolume	0.0640	0.0687
Spread (Δ)	1.1883	1.0654
ϵ-indicator	0.3742	0.1538
Execution time (s)	240 (budget)	240 (budget)

algorithms produce, on average, the same number of non-dominated solutions, MOVNS attains a higher Hypervolume, reflecting a broader coverage of the dominated region of the objective space. At the same time, it yields lower values of Spread and epsilon, which indicates a more uniform distribution of solutions along the frontier and a smaller worst-case distance to the reference set.

Since both approaches were executed under the same computational budget, these differences are due to how each algorithm uses the available time rather than to unequal CPU resources. MOVNS allocates a substantial portion of this budget to intensification and diversification via variable neighborhoods and archive management, whereas NSGA-II relies on population-based exploration guided by crossover, mutation, and crowding distance.

From a practical point of view, MOVNS stands out as a more effective alternative when high-quality and well-distributed sets of trade-off solutions are required, without increasing computational time. NSGA-II, in turn, remains a competitive baseline and offers a simpler implementation and tuning process, but with slightly inferior results in the indicators analyzed here. Overall, the evidence reinforces MOVNS as a robust and attractive option for tackling highly constrained multi-objective problems such as the TTDP.

5   Conclusion

This work presented a *Multi-objective Variable Neighborhood Search* (MOVNS) approach for the *Tourist Trip Design Problem* (TTDP), aiming to generate optimized urban itineraries under multiple conflicting criteria: attractiveness, travel time, estimated cost, and effective visiting time. The metaheuristic was adapted to the specificities of urban tourism, including time windows, multimodal transportation, budget constraints, and user preferences, resulting in a realistic and flexible solution representation.

Under a common time budget of 240 s per run, the experimental evaluation showed that MOVNS outperforms the reference algorithm NSGA-II in the main multi-objective metrics considered, namely Hypervolume, IGD, Spread, and the additive ϵ-indicator, while producing Pareto fronts with a comparable number of non-dominated solutions. The qualitative analysis of itineraries confirmed the ability of the method to generate diverse and personalized alternatives for different traveler profiles, reinforcing the robustness of MOVNS in highly constrained multi-objective settings.

Three aspects of the study are, in our view, the most relevant. First, we integrate real transportation and attraction data into a multi-objective trip planning model. Second, we design neighborhood operators tailored to the tourism domain, with explicit control of temporal and budget feasibility. Third, we incorporate individual preferences into the evaluation of solutions, which allows the itineraries to be aligned with different traveler profiles.

The study has two main limitations. First, the computational cost of MOVNS may become restrictive for much larger instances or stricter real-time requirements. Second, the current model does not account for dynamic events such as weather conditions, congestion, or temporary attraction closures, which can significantly affect urban itineraries.

For future work, we see at least four directions. One is to explore parallelization and adaptive learning strategies to accelerate MOVNS. A second direction is to integrate dynamic contextual data, such as weather or real-time congestion. A third avenue is to extend the comparison to decomposition-based algorithms such as MOEA/D. Finally, it would be useful to study how the model behaves when embedded in interactive recommendation systems for smart urban tourism.

The implementation of this work is publicly available online.[1]

Acknowledgments. This work was supported by the Coordenação de Aperfeiçoamento de Pessoal de Nível Superior Brasil (CAPES) Finance Code 001, the Conselho Nacional de Desenvolvimento Científico e Tecnológico (CNPq), grant (PQ 2021 315059/2021-4), and the Fundação Carlos Chagas Filho de Amparo à Pesquisa do Estado do Rio de Janeiro (FAPERJ), grant (JCNE 2021 E-26/201.451/2022).

References

1. Divsalar, G., Divsalar, A., Jabbarzadeh, A., Sahebi, H.: An optimization approach for green tourist trip design. Soft. Comput. **26**(9), 4303–4332 (2022). https://doi.org/10.1007/s00500-022-06834-1
2. Gavalas, D., Kenteris, M., Konstantopoulos, C., Pantziou, G.: Web application for recommending personalised mobile tourist routes. IET Softw. **6**(4), 313–322 (2012)
3. Kolaee, M.H., Jabbarzadeh, A., Al-e, S.M.J.M., et al.: Sustainable group tourist trip planning: an adaptive large neighborhood search algorithm. Exp. Syst. Appl. **237**, 121375 (2024)
4. Liu, Q., Li, X., Liu, H., Guo, Z.: Multi-objective metaheuristics for discrete optimization problems: a review of the state-of-the-art. Appl. Soft Comput. **93**, 106382 (2020)
5. Pitakaso, R., Srichok, T., Khonjun, S., Gonwirat, S., Nanthasamroeng, N., Boonmee, C.: Multi-objective sustainability tourist trip design: an innovative approach for balancing tourists' preferences with key sustainability considerations. J. Clean. Prod. **449**, 141486 (2024)

---

[1] https://github.com/FilipePessoa30/Montreal-Tour-Planning/tree/main.

6. Raoui, H.E., Cabrera-Cuevas, M., Pelta, D.A.: The role of metaheuristics as solutions generators. Symmetry **13**(11), 2034 (2021)
7. Santos-Peñate, D.R., Moreno-Pérez, J.A., Rodríguez, C.C., Suárez-Vega, R.: A mathematical model and grasp for a tourist trip design problem. In: International Conference on Computer Aided Systems Theory, pp. 112–120. Springer, Cham (2022). https://doi.org/10.1007/978-3-031-25312-6_13
8. Schilde, M., Doerner, K.F., Hartl, R.F., Kiechle, G.: Metaheuristics for the bi-objective orienteering problem. Swarm Intell. **3**(3), 179–201 (2009)
9. Son, N.T., Ha, T.T.N., Jaafar, J.B., Anh, B.N., Giang, T.T.: Some metaheuristics for tourist trip design problem. In: 2023 IEEE Symposium on Industrial Electronics & Applications (ISIEA), pp. 1–10. IEEE (2023). https://doi.org/10.1109/ISIEA58478.2023.10212154
10. Souffriau, W., Vansteenwegen, P., Vertommen, J., Berghe, G.V., Oudheusden, D.V.: A personalized tourist trip design algorithm for mobile tourist guides. Appl. Artif. Intell. **22**(10), 964–985 (2008)
11. Vansteenwegen, P., Souffriau, W., Berghe, G.V., Oudheusden, D.V.: Metaheuristics for tourist trip planning. In: Sörensen, K., Sevaux, M., Habenicht, W., Geiger, M. (eds.) Metaheuristics in the service industry, pp. 15–31. Springer, Heidelberg (2009). https://doi.org/10.1007/978-3-642-00939-6_2
12. Vansteenwegen, P., Souffriau, W., Van Oudheusden, D.: The orienteering problem: a survey. Eur. J. Oper. Res. **209**(1), 1–10 (2011)
13. Wang, Q., Wang, L., Huang, W., Wang, Z., Liu, S., Savić, D.A.: Parameterization of NSGA-II for the optimal design of water distribution systems. Water **11**(5), 971 (2019). https://doi.org/10.3390/w11050971
14. Wisittipanich, W., Boonya, C.: Multi-objective tourist trip design problem in Chiang Mai City **895**(1), 012014 (2020)
15. Zavala, G., Nebro, A.J., Luna, F., Coello Coello, C.A.: Structural design using multi-objective metaheuristics. comparative study and application to a real-world problem. Struct. Multidiscip. Optim. **53**(3), 545–566 (2016)
16. Zeine, A.T., Hami, A.E., Ellaia, R., Pagnacco, E.: Backtracking search algorithm for multi-objective design optimisation. Int. J. Math. Model. Numer. Optim. **8**(2), 93–107 (2017)
17. Zitzler, E., Künzli, S.: Indicator-based selection in multiobjective search. In: Yao, X., et al. (eds.) PPSN 2004. LNCS, vol. 3242, pp. 832–842. Springer, Heidelberg (2004). https://doi.org/10.1007/978-3-540-30217-9_84
18. Zitzler, E., Thiele, L.: Multiobjective evolutionary algorithms: a comparative case study and the strength pareto approach. IEEE Trans. Evol. Comput. **3**(4), 257–271 (1999). https://doi.org/10.1109/4235.797969

ced up to 13-digit precision, yielding improvement of the MIP_mip_gap, which
measures the quality of the resulting feasible integer solution.

Variable Neighborhood Programming for Job Shop Scheduling Problems

Michael-Alexandros Triantoglou[1], Angelo Sifaleras[1(✉)],
and Rachid Benmansour[2]

[1] Department of Applied Informatics, School of Information Sciences, University of
Macedonia, 156 Egnatias Street, 54636 Thessaloniki, Greece
mtriantoglou@uom.edu.gr, sifalera@uom.gr
[2] Research Laboratory in Information Systems, Intelligent Systems and Mathematical
Modeling, National Institute of Statistics and Applied Economics, Rabat, Morocco
r.benmansour@insea.ac.ma

Abstract. The Job Shop Scheduling Problem is a classic combinatorial optimization problem and one of the most well-studied scheduling problems. Several methodologies, both exact and metaheuristic, have already been proposed for the solution of this computationally difficult problem. This work presents for the first time a solution approach based on Variable Neighborhood Programming for the Job Shop Scheduling Problem. Variable Neighborhood Programming is a recent methodology which constitutes a combination of Genetic Programming and Variable Neighborhood Search. In addition, some encouraging comparative computational results are also shown against the state-of-the-art Gurobi optimization solver using medium- and large-scale benchmark instances. The findings of this work have a plethora of modern applications in Manufacturing-as-a-Service online platforms. All experimental evaluations were performed on the Google Cloud Platform.

Keywords: Variable Neighborhood Programming · Job Shop
Scheduling Problem · Mixed-Integer Programming ·
Manufacturing-as-a-Service · Google Cloud Platform

1 Introduction

Scheduling problems constitute a large area of optimization problems [1]. The Job Shop Scheduling Problem (JSSP) is one of the most well-studied scheduling problems [14]. Thus, several researchers have tried various solution approaches including exact methodologies, metaheuristic, Constraint Programming (CP), and others [26].

In JSSP, there are a set of jobs and a set of machines. Each job consists of a sequence of activities (otherwise, tasks or operations). Each activity must be processed on a specific machine and has a fixed processing time. Activities of the same job must follow the given order, and each machine can process at most one

S. Cavero et al. (Eds.): ICVNS 2025, LNCS 16256, pp. 120–133, 2026.
https://doi.org/10.1007/978-3-032-19582-1_9

operation at a time. A schedule assigns starting times to all activities so that the activities on the same machine do not overlap. The goal is usually to finish all jobs as early as possible, which means minimizing the total makespan subject to precedence and machine capacity constraints [18].

The JSSP is also $\mathcal{NP}$-hard since it consists of a generalization of the $\mathcal{NP}$-hard Traveling Salesman Problem (TSP), if we consider a single job (as a TSP salesman) and machines (as TSP cities). Therefore, finding the exact optimal solution becomes very difficult when the problem size increases. For this reason, several researchers have developed parallel implementations that significantly improve performance. Taillard presented a parallel Taboo Search metaheuristic for the solution of the JSSP [22]. Also, machine learning techniques are now widely used for the solution of the JSSP. Zhang & Zhu recently compiled a survey of reinforcement learning methods applied to the JSSP [28].

Scheduling problems also arise in modern Manufacturing-as-a-Service (MaaS) online platforms. Such distributed manufacturing ecosystems usually provide a decomposition of scheduling requests and a composition of supply chains using advanced artificial intelligence and optimization techniques. Recently, Lagos et al. proposed a CP model for the efficient solution of flow shop scheduling problems with multiple objectives in the plastic injection molding domain [13]. The research methodology of this paper will also be applied in similar real-world optimization problems that arise on the Tec4MaaSEs platform http://tec4maases.eu.

The research contribution of this paper is the development, for the first time, of a Variable Neighborhood Programming method for the efficient solution of large-scale JSSP instances. The remainder of this paper is organized as follows. Section 2 gives the mathematical formulation of the JSSP. In Sect. 3, we present the research methodology and an illustrative example. The proposed VNP approach was tested using well-known benchmark results against the state-of-the-art Gurobi optimization solver, and the experimental results are shown in Sect. 4. Finally, conclusions and future work are discussed in Sect. 5.

2 Mathematical Formulation of the JSSP

Several mathematical formulations have already been proposed for the solution of the JSSP, such as the time-indexed formulation [5], the rank-based formulation [25], and the disjunctive formulation [14]. This section will provide the disjunctive formulation proposed by Mann [14] in 1960, which is considered the best performing Mixed Integer Programming (MIP) model for the JSSP according to theoretical [17] and experimental comparisons [12] in the literature.

Indices and Sets

- $i, k \in \{1, \ldots, n\}$: Index for jobs
- $j \in \{1, \ldots, m\}$: Index for machines

Parameters

- t_{ij}: Processing time of job i on machine j.
- M: A large positive number (big-M).

Decision Variables

- x_{ij}: The start time of the job i on machine j.
- y_{ikj}: A binary variable, where $y_{ikj} = 1$ if job i precedes job k on machine j, and 0 otherwise.
- $C_{\max}$: The makespan (i.e., the completion time of the last job).

Mathematical Formulation

The objective is to minimize the makespan:

$$\text{Minimize} \quad C_{\max} \tag{1}$$

Subject to the following constraints:

$$x_{i,j} + t_{ij} \leq x_{i,j+1} \qquad \forall i \in \{1,\ldots,n\}, \quad \forall j \in \{1,\ldots,m-1\} \tag{2}$$

$$x_{ij} + t_{ij} \leq x_{kj} + M \cdot (1 - y_{ikj}) \quad \forall j \in \{1,\ldots,m\}, \quad \forall i,k \in \{1,\ldots,n\}, i < k \tag{3}$$

$$x_{kj} + t_{kj} \leq x_{ij} + M \cdot y_{ikj} \qquad \forall j \in \{1,\ldots,m\}, \quad \forall i,k \in \{1,\ldots,n\}, i < k \tag{4}$$

$$x_{in} + t_{in} \leq C_{\max} \qquad \forall i \in \{1,\ldots,n\} \tag{5}$$

$$x_{ij} \geq 0 \qquad \forall i \in \{1,\ldots,n\}, \quad \forall j \in \{1,\ldots,m\} \tag{6}$$

$$y_{ikj} \in \{0,1\} \qquad \forall j \in \{1,\ldots,m\}, \quad \forall i,k \in \{1,\ldots,n\}, i < k \tag{7}$$

The precedence constraints (2) ensure that for any given job, the activities are performed in the correct sequence. This formulation assumes that the technological order of the machines is $1, 2, \ldots, n$ for all jobs. The pair of disjunctive constraints (3) and (4) ensures that no two jobs can be processed on the same machine simultaneously. If $y_{ikj} = 1$, job i must finish before job k starts on machine j (constraint (4) becomes redundant). If $y_{ikj} = 0$, job k must be completed before job i starts (constraint (3) becomes redundant). In addition, the constraint (5) ensures that the makespan must be greater than or equal to the completion time of the last operation of every job. Finally, the constraints (6) and (7) define the domains of the decision variables.

3   Research Methodology

Variable Neighborhood Programming (VNP) is a more recent variant of the VNS metaheuristic framework [10,15] for Automatic Programming (AP). The idea is to treat rules (expressed as AP trees) as programs and then systematically modify them by exploring different neighborhoods. Essentially, it is a combination of Variable Neighborhood Search (VNS) and Genetic Programming (GP). GP is an evolutionary algorithm [19] with several applications in scheduling [27]. Genetic Programming works with "AP trees" that represent small programs or rules. These AP trees have operators (for example $+$, $-$, $\times$, min, max) and terminals, which are the problem's features, such as the processing time or remaining operations.

One simple way to build schedules is by using dispatching rules [23], which are simple heuristics that decide which job should be processed next when a machine becomes free. Examples include FIFO (First In First Out) where we process jobs in the order they arrive, SPT (Shortest Processing Time) where we choose the job with the shortest processing time, LPT (Longest Processing Time) where we choose the job with the longest processing time, MWKR (Most Work Remaining) where we choose the job with the largest amount of work left, EDD (Earliest Due Date), and others. These rules are often used as constructive heuristic methods. In [16]. The SPT rule is one of the oldest and most important. In 1956, Smith proved that ordering jobs by increasing processing time is the best way to minimize the total completion time for a single machine problem [20]. Because of this, SPT is often used as an initial solution in more advanced methods. Dispatching rules can be expressed as priority functions based on job and operation attributes, also called terminals. Commonly used attributes are the following [27]:

- pt: processing time of the current activity
- rt: total remaining processing time of the job
- ro: number of remaining operations of the job
- dd: due date of the job
- wt: waiting time before the activity can start
- idx: index of the current activity in its job sequence (shows the position of the activity inside the job)

By combining these attributes with the operators described above (i.e., $+$, $-$, $\times$, min, max), more complex rules can be created. For example, the expression $\min(pt \times 1.0, ro \times 1.0) + rt \times 1.0$ is a composite rule that can be used to assign priorities. The interval $[-1, 1]$ is often used for the coefficient values at the terminal nodes [15]. This choice makes the search process simpler and faster while still giving enough flexibility to adjust the solution. Using this limited range has been reported to improve convergence speed compared to larger intervals, which is why it is considered a suitable option in practice.

In this work, the SPT priority rule, according to which jobs are executed in order of increasing processing time, is used to initialize the decision AP tree in the

VNP. This rule [20] was first presented by Smith in 1956 as an optimal strategy for minimizing the mean flow time in single-stage scheduling environments.

VNS is a strategy that improves solutions by systematically applying different types of changes, called "neighborhoods". If one neighborhood does not give an improvement, the search moves to the next one, in order to avoid getting stuck in local optima. The VNS metaheuristic also finds a plethora of applications in various fields [4,6] and also in scheduling [2,3,7].

In the VNP context, a neighborhood is defined by a small change in the rule structure, such as changing a weight, replacing an operator, or adding a new subtree. By moving from smaller to larger neighborhoods, VNP can escape local optima and generate more effective dispatching rules. In this way, VNP is used to evolve scheduling heuristics for the JSSP with the objective of minimizing the makespan. The starting point is an initial rule (for example, SPT), and then various neighborhoods can be applied.

3.1 Solution Representation

The solution representation, (i.e., AP tree) of this problem is shown below in Fig. 1.

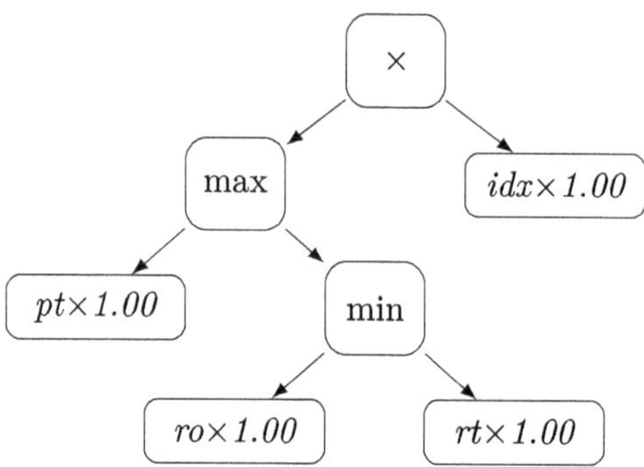

Fig. 1. Solution representation using AP tree.

At the root there is a × operator, which multiplies two branches. On the left branch, there is a *max* operator that compares two values. The first value is obtained by multiplying *pt* by the weight 1.00, while the second value comes from another activity: a *min* between $ro \times 1.00$ and $rt \times 1.00$. On the right branch, the value is simply $idx \times 1.00$. Therefore, the solution corresponds to the following function:

$$(\max\left(\left(pt \times 1.00\right), \min\left(\left(ro \times 1.00\right), \left(rt \times 1.00\right)\right)\right)) \times (idx \times 1.00)$$

3.2   Variable Neighborhood Programming

In the proposed study, a General VNP scheme was used in a manner similar to that of General VNS. Thus, the General VNP consists of the exploitation phase (i.e., Variable Neighborhood Descent - VND) and the exploration phase (i.e., shaking process). The proposed VNP algorithm employs the set $\mathcal{N}$ of the following four neighborhoods that are applied to the rule tree.

- N_1: change terminal/weight (mutate terminal),
- N_2: replace operator,
- N_3: add subtree, and
- N_4: replaces subtree.

These four neighborhoods correspond to classic GP mutations (terminal/function/subtree mutation), as described by Koza in [11]. In addition, recent work on VNP study specific neighborhood structures for AP trees [9]. The first three neighborhoods as shown in Fig. 2 are used in the intensification phase (i.e., in the VND part). The fourth neighborhood as shown in Fig. 3 is used in the diversification phase of the VNP (i.e., the shake procedure of the VNP).

Since three neighborhoods are used during the local search, the cardinality number of the neighborhoods (i.e., l_{max}) will be equal to three. The VND algorithm is shown below in the pseudocode 1. The neighborhood structure is sequentially changed at each iteration once there is no other improvement using the current neighborhood structure.

Algorithm 1. VND

1: **Input:** $T, l_{max}, \mathcal{N}$
2: **Output:** T'
3: **repeat**
4:     stop $\leftarrow$ **true**
5:     $l \leftarrow 1$
6:     **while** $l \leq l_{max}$ **do**
7:         $T' \leftarrow \arg\min_{y \in \mathcal{N}_l(T)} f(y)$ {Find the best makespan in the current neighborhood}
8:         **if** $f(T') \leq f(T)$ **then**
9:             $T \leftarrow T'$ {make a move}
10:             $l \leftarrow 1$
11:             stop $\leftarrow$ **false**
12:         **else**
13:             $l \leftarrow l + 1$ {next neighborhood}
14:         **end if**
15:     **end while**
16: **until** stop = **true**
17: **return** T'

The shaking process in VNP follows the same idea as in the VNS method. It is an adaptive shaking using a different neighborhood (N_4) that replaces subtrees

(a) The first neighborhood operator (*mutate_terminal*) changes the weight of the initial node.

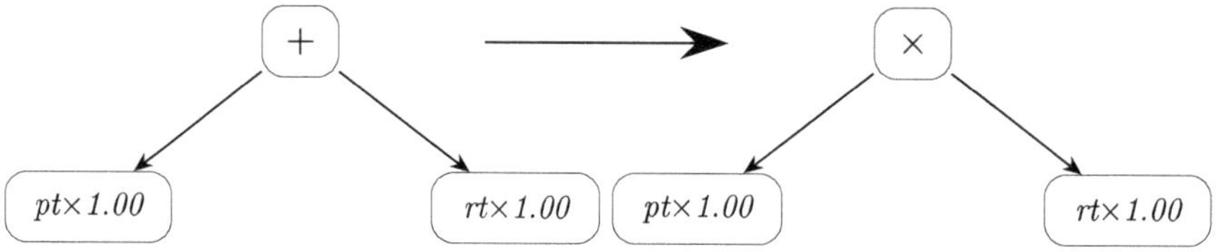

(b) The second neighborhood operator (*replace_operator*) changes the operator of the root from $+$ into $\times$.

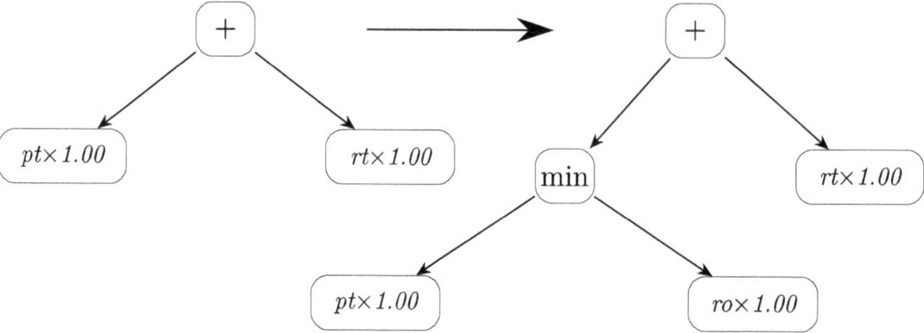

(c) The third neighborhood operator (*add_subtree*) adds a new left subtree $\min(pt, ro)$.

Fig. 2. The three neighborhoods used for local search in the VND algorithm.

for scheduling rules based on GP. The intensity level of the shaking process (that is, how many times we repeat a non-improved solution to escape from the local optimum) is denoted by k_{max}. Its purpose is to escape from the current local optimum by applying a different neighborhood, which randomly changes the priority rule tree. In the present implementation, shaking allows replacing parts of the AP tree with randomly generated structures, while keeping a restriction on the AP tree size to avoid excessive growth in complexity and k_{max} was set equal to three. In this way, VNP avoids premature convergence and explores new areas of the solution space for the JSSP. The shaking procedure is shown below in the pseudocode 2.

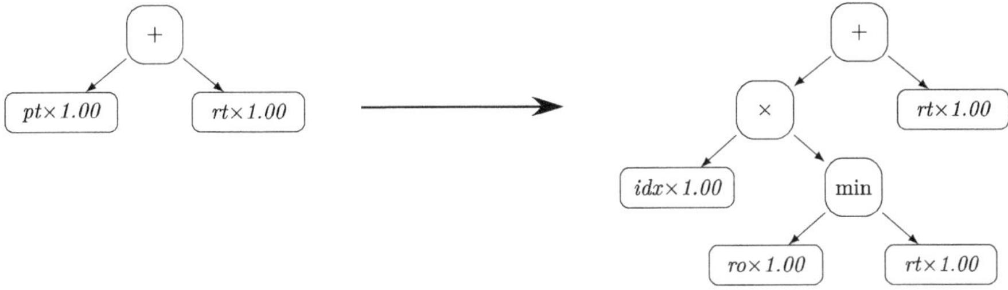

Fig. 3. The fourth neighborhood operator (*replace_subtree*) which replaces a subtree, is used in the shaking phase.

Algorithm 2. Shake

1: **Input:** T, k
2: **Output:** T'
3: **for** $i \leftarrow 1$ **to** k **do**
4:     Choose randomly $T' \in N_4(T)$
5:     $T \leftarrow T'$
6: **end for**
7: **return** T

The complete General VNP algorithm is shown below in the pseudocode 3.

Algorithm 3. General VNP for JSSP

1: **Input:** jobs, machines, activities, $k_{max}, l_{max}, \mathcal{N}$
2: **Output:** schedule, makespan
3: {Initialization}
4: Create an AP tree solution (T) using the SPT rule
5: **repeat**
6:   $k \leftarrow 1$
7:   **while** $k \leq k_{max}$ **do**
8:       $T' \leftarrow$ Shake(T, k)
9:       $T'' \leftarrow$ VND$(T', l_{max}, \mathcal{N})$
10:   **end while**
11: **until** maximum CPU time limit

3.3 Illustrative Example

To better understand how the three neighborhoods work, let us assume the following small 3×3 instance in the common Taillard format [21]:

```
3 3
0 3 1 2 2 2
1 2 2 1 0 4
2 4 0 3 1 1
```

As can be seen from the header line, there are three jobs and three machines. Each job consists of ordered sequences of activities in the following lines. Thus, according to the first row (Job 0), the first activity must be on machine 0 and takes 3 units of time, the second activity must be on machine 1 and takes 2 units of time, and finally the third activity must be on machine 2 and takes 2 units of time.

Figure 4a shows the initial solution (i.e., terminal node). The initial solution is generated using the dispatching rule: Processing time (pt) × 1.00. This means that the priority of each activity is determined solely by its processing time.

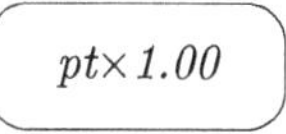

(a) At first, SPT generates an initial AP tree solution (root terminal), which corresponds to a pt node with weight 1 and a makespan of 15.

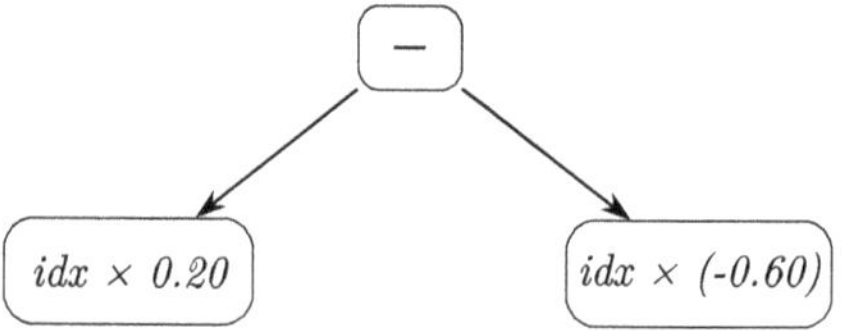

(b) Final AP tree solution with a makespan of 11.

Fig. 4. Initial and final AP tree solutions.

Figure 4b shows the final solution after several iterations for this specific example. The final solution is generated using the dispatching rule: (idx × 0.20) − (idx × (−0.60)) = 0.80 × idx. Therefore, the priority score is simply: Priority = 0.80 × idx. We use the rule $Priority = 0.80 × idx$ and always pick the job with the smallest priority. Because 0.80 is positive, this is the same as sorting by idx: all first activities ($idx = 0$) are chosen before any second ones ($idx = 1$), then the third ones ($idx = 2$), and so on. If two jobs have the same idx, we break ties by the job's index (lower id first). After choosing an activity, we place it at the earliest time when both its job and the required machine are free. In short, the schedule is built layer by layer across stages: the first stage of every job, then the second stage of every job, then third, etc.

Figure 5 shows the two Gantt charts corresponding to the initial and final solutions, respectively. As can be seen, the initial makespan is equal to 15 and the final makespan was equal to 11.

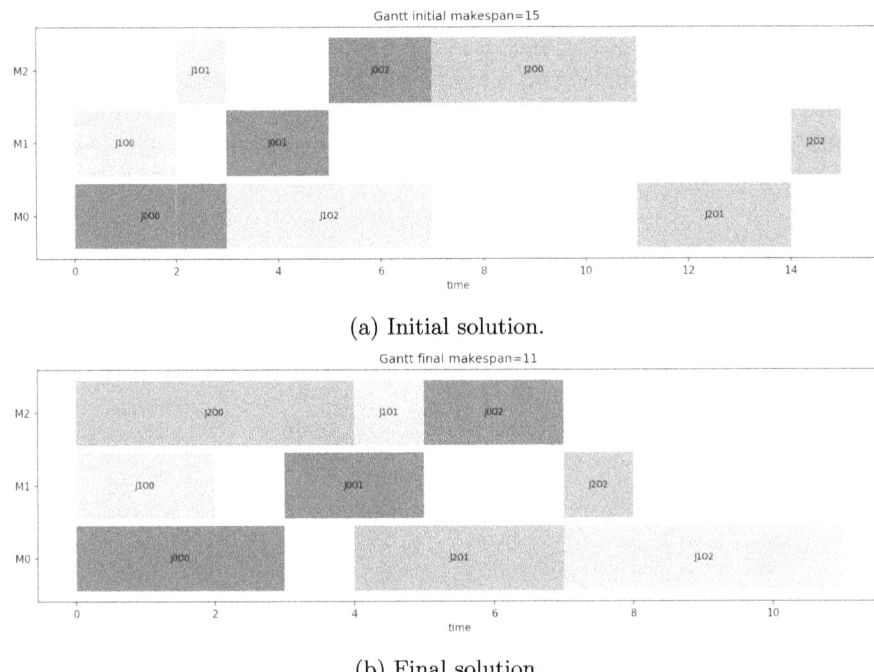

(a) Initial solution.

(b) Final solution.

Fig. 5. Initial and final solutions.

4   Experimental Results and Comparison

The computational experiments were performed on a virtual machine on the Google Cloud Platform running 64-bit Windows Server 2022 Datacenter with an Intel Xeon CPU at 3.10 GHz featuring 1 socket with 8 cores and 16 threads (2 threads per core) and 64 GB main memory. Also, both VNP and Gurobi were implemented in Python 3.13.7.

Efforts were made to solve each one of them using the latest version of the state-of-the-art Gurobi optimizer v12.0.3 within a reasonable amount of time set to 10 min. The time limit (stopping condition) of our VNP approach was 30 s. However, due to the increased computational difficulty, Gurobi was unable to solve any instance to optimality. Thus, Gurobi has only computed an upper bound of the optimal objective value. Therefore, Table 1 reports the comparative numerical results regarding the incumbent solutions found by Gurobi (time limit = 600 s) and the proposed implementation of the VNP (time limit = 30 s).

Each implementation was tested using well-known benchmark instances with dimensions ranging from 20 to 50 machines and jobs ranging from 15 to 100. The data set is publicly available online at https://www.jobshoppuzzle.com/benchmarks.html. These instances belong to the standard general JSSP and not in any other special case; thus, the machine order is not the same for all jobs.

Each instance was solved 10 times using the VNP method, and the results of the comparative computational study are shown in Table 1, where the best values per instance are indicated in bold font. The name and dimension of each instance are depicted in the first two columns. Columns three and four show the average makespan computed by the proposed VNP method and the makespan found by the Gurobi optimizer, respectively.

Table 1. Comparative computational study VNP Vs Gurobi

Instance	Dimension	Gurobi Makespan (600 s)	VNP Makespan (30 s)
Ta54	50×15	3,354	3,303.9
Ta55	50×15	3,498	3,497.0
Ta56	50×15	3,690	3,348.2
Ta60	50×15	3,265	3,198.4
Ta61	50×20	4,098	3,659.2
Ta62	50×20	3,949	3,792.8
Ta65	50×20	4,269	3,557.5
Ta66	50×20	3,704	3,573.9
Ta67	50×20	3,996	3,662.1
Ta70	50×20	4,629	3,899.8
Ta71	100×20	69,198	6,900.1
Ta73	100×20	68,300	6,656.5
Ta74	100×20	69,133	6,103.4
Ta75	100×20	57,903	6,571.0
Ta76	100×20	69,335	6,406.4
Ta77	100×20	78,561	6,249.0
Ta80	100×20	72,573	6,203.6

In addition, a maximum depth of three levels was used for any new subtree in the proposed implementation, and also a size limit of 200 nodes was set for the entire tree. In case the tree exceeds this limit, then it is pruned by replacing a subtree with a terminal node.

Figure 6 shows the percentage improvement (%) of the proposed VNP approach compared to the Gurobi optimizer. As can be seen, the computational benefits of VNP increase as the dimension of the problem increases. Although only the Gurobi solver utilized the 16 threads, since the VNP code was not parallelized, VNP found better solutions than Gurobi in a short amount of time.

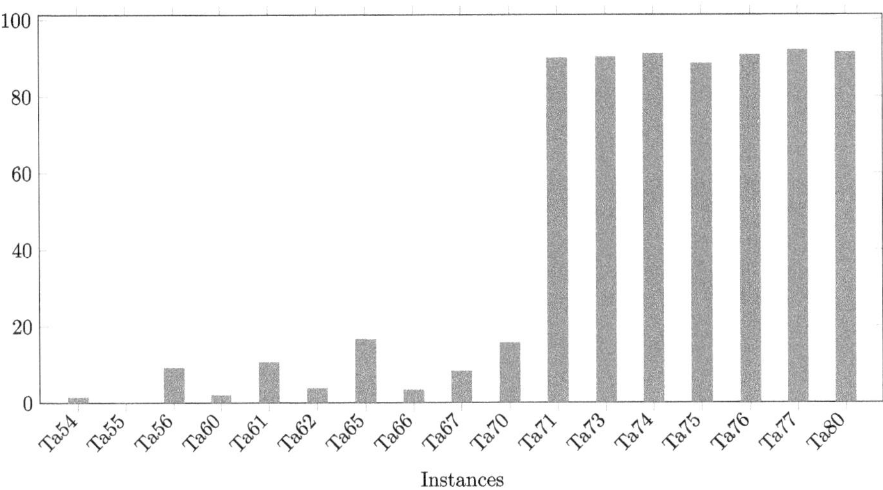

Fig. 6. Improvement of VNP over Gurobi (%).

4.1   Discussion

The JSSP is $\mathcal{NP}$-hard with a massive search space, and the classical disjunctive model has a weak linear relaxation, making branch-and-bound inefficient. Therefore, Gurobi often finds it difficult to obtain good incumbent solutions within the time limit. From the above Table 1 it can be seen that the VNP compared to Gurobi found a smaller makespan in all these instances. Also, in some larger instances with 20 machines and 100 jobs, VNP in just a few seconds outperformed Gurobi that was running with 16 threads even in 600 s and found a much better solution. The largest difference is in large instances, where VNP gives up to 90% better solutions. This shows very clearly that VNP works better when the problem size grows, and thus constitutes a very competitive method for other complex scheduling problems.

5   Conclusions and Future Work

The results show that the VNP approach is promising for the JSSP. The proposed VNP method outperformed the Gurobi optimizer in situations requiring a quick solution. Therefore, the proposed VNP solution method can be a useful tool for managers in competitive real-world Manufacturing-as-a-Service ecosystems such as the Tec4MaaSEs online platform.

For future research, it is suggested to investigate more neighborhoods within the VNP framework and try to apply the method to larger instances of JSSP [24]. Also, a computational comparison with other efficient solution methods, such as CP, would be very interesting. In addition, other researchers have suggested more advanced operators, such as the Elementary Tree Transformation Local Search

(ETTLS) [8]. Adding ETTLS and comparing it with the proposed method could also be a future extension of this study. Finally, another direction is to analyze the performance of commercial solvers when combined with VNP as a warm-start strategy.

Acknowledgements. This research was funded by the European Health and Digital Executive Agency, Project: 101138517, Tec4MaaSEs, HORIZON-CL4-2023-TWIN-TRANSITION-01.

Google Cloud Platform (GCP) resources were provided by the Greek National Infrastructures for Research and Technology GRNET and funded by the EU Recovery and Resiliency Facility.

References

1. Agnetis, A., Billaut, J.C., Pinedo, M., Shabtay, D.: Fifty years of research in scheduling — theory and applications. Eur. J. Oper. Res. **327**(2), 367–393 (2025)
2. Benbrik, O., Benmansour, R., Elidrissi, A., Sifaleras, A.: Advanced algorithms for the reclaimer scheduling problem with sequence-dependent setup times and availability constraints. In: Sevaux, M., Olteanu, AL., Pardo, E.G., Sifaleras, A., Makboul, S. (eds.) MIC 2024. LNCS, vol. 14753, pp. 291–308. Springer, Cham (2024). https://doi.org/10.1007/978-3-031-62912-9_28
3. Benmansour, R., Sifaleras, A.: Scheduling in parallel machines with two servers: the restrictive case. In: Mladenovic, N., Sleptchenko, A., Sifaleras, A., Omar, M. (eds.) ICVNS 2021. LNCS, vol. 12559, pp. 71–82. Springer, Cham (2021). https://doi.org/10.1007/978-3-030-69625-2_6
4. Sifaleras, A., Salhi, S., Brimberg, J. (eds.): ICVNS 2018. LNCS, vol. 11328. Springer, Cham (2019). https://doi.org/10.1007/978-3-030-15843-9
5. Bowman, E.H.: The schedule-sequencing problem. Oper. Res. **7**(5), 621–624 (1959)
6. Brimberg, J., Salhi, S., Todosijević, R., Urošević, D.: Variable neighborhood search: the power of change and simplicity. Comput. Oper. Res. **155**, 106221 (2023)
7. Elidrissi, A., Benmansour, R., Sifaleras, A.: A general variable neighborhood search for the parallel machine scheduling problem with two common servers. Optim. Lett. **17**(9), 2201–2231 (2023)
8. Elleuch, S., Hansen, P., Jarboui, B., Mladenović, N.: New VNP for automatic programming. Electron. Notes Discrete Math. **58**, 191–198 (2017)
9. Elleuch, S., Jarboui, B.: Analysis of six different GP-tree neighborhood structures. In: Abraham, A., Gandhi, N., Hanne, T., Hong, T.-P., Nogueira Rios, T., Ding, W. (eds.) ISDA 2021. LNNS, vol. 418, pp. 1237–1249. Springer, Cham (2022). https://doi.org/10.1007/978-3-030-96308-8_115
10. Elleuch, S., Jarboui, B., Mladenović, N., Pei, J.: Variable neighborhood programming for symbolic regression. Optim. Lett. **16**(1), 191–210 (2022)
11. Koza, J.R.: Genetic Programming: On the Programming of Computers by Means of Natural Selection. MIT Press, Cambridge (1992)
12. Ku, W.Y., Beck, J.C.: Mixed integer programming models for job shop scheduling: a computational analysis. Comput. Oper. Res. **73**, 165–173 (2016)
13. Lagos, A.I., Avgerinos, I., Zois, G., Mourtos, I., Casla, P.: Optimising a manufacturing-as-a-service platform through mathematical modeling. In: Mizuyama, H., Morinaga, E., Nonaka, T., Kaihara, T., von Cieminski, G., Romero,

D. (eds.) APMS 2025. IFIP, vol. 769, pp. 478–493. Springer, Cham (2025). https://doi.org/10.1007/978-3-032-03550-9_32

14. Manne, A.S.: On the job-shop scheduling problem. Oper. Res. **8**(2), 219–223 (1960)

15. Mladenovic, N., Jarboui, B., Elleuch, S., Mussabayev, R., Rusetskaya, O.: Variable neighborhood programming as a tool of machine learning. In: Pardalos, P.M., Rasskazova, V., Vrahatis, M.N. (eds.) Black Box Optimization, Machine Learning, and No-Free Lunch Theorems. SOIA, vol. 170, pp. 221–271. Springer, Cham (2021). https://doi.org/10.1007/978-3-030-66515-9_9

16. Nguyen, S., Zhang, M., Johnston, M., Tan, K.C.: Genetic programming for job shop scheduling. In: Bansal, J.C., Singh, P.K., Pal, N.R. (eds.) Evolutionary and Swarm Intelligence Algorithms. SCI, vol. 779, pp. 143–167. Springer, Cham (2019). https://doi.org/10.1007/978-3-319-91341-4_8

17. Pan, C.H.: A study of integer programming formulations for scheduling problems. Int. J. Syst. Sci. **28**(1), 33–41 (1997)

18. Pinedo, M.L.: Scheduling: Theory, Algorithms, and Systems, 6th edn. Springer, Cham (2022). https://doi.org/10.1007/978-3-031-05921-6

19. Sette, S., Boullart, L.: Genetic programming: principles and applications. Eng. Appl. Artif. Intell. **14**(6), 727–736 (2001)

20. Smith, W.E.: Various optimizers for single-stage production. Naval Res. Logistics Q. **3**(1–2), 59–66 (1956)

21. Taillard, E.: Benchmarks for basic scheduling problems. Eur. J. Oper. Res. **64**(2), 278–285 (1993)

22. Taillard, E.D.: Parallel taboo search techniques for the job shop scheduling problem. ORSA J. Comput. **6**(2), 108–117 (1994)

23. Đurasević, M., Jakobović, D.: Evolving dispatching rules for optimising many-objective criteria in the unrelated machines environment. Genet. Program Evolvable Mach. **19**(1), 9–51 (2018)

24. Van Hoorn, J.J.: The current state of bounds on benchmark instances of the job-shop scheduling problem. J. Sched. **21**(1), 127–128 (2018)

25. Wagner, H.M.: An integer linear-programming model for machine scheduling. Naval Res. Logistics Q. **6**(2), 131–140 (1959)

26. Xiong, H., Shi, S., Ren, D., Hu, J.: A survey of job shop scheduling problem: the types and models. Comput. Oper. Res. **142**, 105731 (2022)

27. Zhang, F., Nguyen, S., Mei, Y., Zhang, M.: Genetic Programming for Production Scheduling. Springer, Cham (2021). https://doi.org/10.1007/978-981-16-4859-5

28. Zhang, X., Zhu, G.Y.: A literature review of reinforcement learning methods applied to job-shop scheduling problems. Comput. Oper. Res. **175**, 106929 (2025)

Variable Formulation Search
for the Cyclic Min-Max Sitting
Arrangement Problem

Marcos Robles[✉][iD], Sergio Cavero[iD], and Eduardo G. Pardo[iD]

Universidad Rey Juan Carlos, Madrid, Spain
{marcos.robles,sergio.cavero,eduardo.pardo}@urjc.es

Abstract. The *Cyclic Min-Max Sitting Arrangement* (CMMSA) is a
graph layout optimization problem where the vertices of an input signed
graph must be assigned to those of a cyclic host graph in a one-to-one
correspondence. The input graph contains weighted edges, $+1$ or -1,
representing positive and negative relationships between the vertices.
For each vertex, a penalty occurs when an adjacent vertex connected by
a negative-labeled edge is positioned closer in the cycle than any other
adjacent vertex connected by a positive-labeled edge. The goal is to minimize the maximum number of such penalties occurring at any single vertex. This paper presents a comprehensive study based on the *Variable
Neighborhood Search* (VNS) methodology for solving the CMMSA. Building upon successful applications of VNS in related problems, we analyze
multiple algorithmic variants and strategies. Specifically, our investiga-
tion focuses on *Variable Formulation Search* (VFS), which defines multiple formulations for the problem, allowing it to further explore the solution space. We propose a total of four alternative formulations specially
suited for this problem. The use of VFS within the VNS methodology
outperforms the use of a default VNS schema, obtaining a better solution in 19 out of 20 instances considered in the study. The findings not
only highlight the efficacy of the proposed approach but also provide a
foundation for future research on the CMMSA and related graph layout
optimization problems.

Keywords: Sitting Arrangement · Cyclic Graph Layout Problem ·
Signed graph · Variable Formulation Search

1 Introduction

Graph Layout Problems (GLPs) constitute a family of combinatorial optimization problems aimed at establishing relationships between vertices of an input
graph and vertices of a host graph in such a way that optimizes a given objective
function [5]. The initial practical application of these problems emerged in cir-
cuit design, where given a set of components, the goal was to distribute them on
a board to minimize aspects such as wiring requirements or cable crossings [10],

S. Cavero et al. (Eds.): ICVNS 2025, LNCS 16256, pp. 134–149, 2026.
https://doi.org/10.1007/978-3-032-19582-1_10

but they have been found to have real-world applications in other fields such as numerical analysis, information retrieval, and scheduling.

Within the GLP family, there exists a subset of problems denoted as *Sitting Arrangement* (SA) problems, which were introduced in 2011 [7]. These problems enable the representation of social situations by indicating positive and negative relationships between the vertices, making them particularly compelling, as they can model scenarios that cannot be replicated in other problems that employ simple graphs. Similar to other GLP problems, SA problems are defined through a combination of four elements: the input graph, the host graph, an embedding function, and the objective function to be optimized.

The input graph in SA problems is characterized as a signed graph that is finite, undirected, and features labels of either $+1$ or -1 on its edges [17]. Formally, it is defined as $G = (V_G, E_G, \sigma)$, where V_G and E_G represent the vertex set and edge set, respectively. The function σ is associated with graph G, and returns the label $+1$ or -1 for a given edge, that is, $\sigma : E_G \to \{-1, +1\}$. Additionally, we define the function $\Gamma_G(u)$ which, given a vertex $u \in V_G$, returns all vertices $v \in V_G$ such that $\{u, v\} \in E_G$. Analogously, this function can be extended to $\Gamma_G^+(u)$ and $\Gamma_G^-(u)$, which would return the sets of vertices adjacent to u linked by positive or negative labeled edges, respectively. For the sake of simplicity, these vertices will be referred to as "positive adjacents" and "negative adjacents" throughout this article.

Figure 1a presents an example of a signed graph G_1 containing five vertices and five edges. The vertex set comprises $V_{G_1} = \{A, B, C, D, E\}$, and the edge set is given by $E_{G_1} = \{\{A, B\}, \{A, C\}, \{A, D\}, \{A, E\}, \{D, E\}\}$, such that $\sigma(\{A, B\}) = -1$, $\sigma(\{A, C\}) = +1$, $\sigma(\{A, D\}) = +1$, $\sigma(\{A, E\} = -1$, and $\sigma(\{D, E\}) = -1$.

Regarding the host graph, SA problems were defined for a path-type host graph [12], and later extended to cycle-type host graphs [1, 13], which is the focus of this paper. A cycle graph is defined as a connected, finite, undirected, and simple graph with a regular structure where each vertex has degree exactly two. This graph is defined as $H = (V_H, E_H)$, where V_H and E_H are the vertex and edge sets of the host graph, respectively.

The path function $\Psi(i, j)$ on the host graph is defined in such a way that given two vertices $i, j \in V_H$, it returns two sets of connected vertices forming a path from i to j. Note that in a cycle, there are two paths between any pair of vertices i, j, with $i \neq j$, without considering loops. Additionally, we also define the function $\psi(i, j)$, which selects the shortest path (i.e., the one composed of the fewest vertices) between them. In the special case where both paths have equal length, the one that optimizes the objective function of the problem is chosen. If a tie still persists, a path is chosen at random.

Figure 1b illustrates an example of a cycle graph H_1 with five vertices. In this case, the vertex set consists of $V_{H_1} = \{1, 2, 3, 4, 5\}$ and the edge set is $E_{H_1} = \{\{1, 2\}, \{2, 3\}, \{3, 4\}, \{4, 5\}, \{5, 1\}\}$. To exemplify the concept of paths, we evaluate the function $\Psi(1, 3)$, which yields $(\{1, 2, 3\}, \{1, 5, 4, 3\})$, each corresponding to a possible path in the cycle. Here, the first path has a length of three, while the second has a length of four. Therefore, if we were to evaluate $\psi(1,3)$, it would return the first path as it is the shortest.

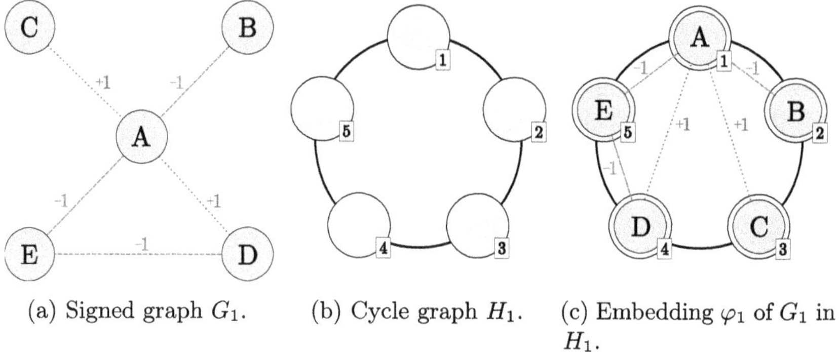

(a) Signed graph G_1.   (b) Cycle graph H_1.   (c) Embedding φ_1 of G_1 in H_1.

Fig. 1. Example of an embedding of an input graph on a cycle graph.

In this work we study a variant of the SA problem known as *Cyclic Min-Max Sitting Arrangement* (CMMSA) [13]. In the CMMSA, as any other problem belonging to the GLP family, the vertices of the input graph are related to the vertices of the host graph through an embedding function such as $\varphi : v \rightarrow i$, where $v \in V_G$ and $i \in V_H$. This bijective function represents a solution and is therefore evaluated with the objective function to quantify its quality. The general objective of CMMSA consists of finding the embedding or solution φ among all possible Φ that minimizes its objective function. Specifically, the objective function of the CMMSA counts, for each input vertex $u \in V_G$, the number of times that any negative adjacents $w \in \Gamma^-(u)$ is assigned to a host vertex $\varphi(w)$ which is found on the shortest path $\psi(\varphi(u), \varphi(v))$ to any positive adjacent $v \in \Gamma^+(u)$. This specific situation has been denoted in the literature as an "error" or "conflict". Formally, the conflict count for a vertex, $\epsilon(u, \varphi)$, in the embedding is defined as follows:

$$\epsilon(u, \varphi) = \sum_{v \in \Gamma_G^+(u)} \left| \{ \, w \in \Gamma_G^-(u) : \varphi(w) \in \psi(\varphi(u), \varphi(v)) \} \right|,$$

where $\varphi(w) \in \psi(\varphi(u), \varphi(v))$ represents the condition that negative adjacent w is found in the path to positive adjacent v.

In the original *Cyclic Minimum Sitting Arrangement* (CMinSA) variant [13], the objective function is obtained by adding individual conflicts $\epsilon(u, \varphi)$. However, this function presents a drawback: the individual evaluation value for a particular vertex might be specially harmful, and yet the solution can still be considered of high quality. However, there are practical situations where it is necessary to find a balance among the vertices with conflicts. To address these situations, this work studies the *Cyclic Min-Max Sitting Arrangement* (CMMSA) variant, which instead of evaluating the objective function using the sum, employs the maximum. This problem is formalized as follows:

$$\mathrm{CMMSA}(\varphi) = \max_{u \in V_G} \epsilon(u, \varphi).$$

The previous definitions are exemplified using the embedding φ_1 shown in Fig. 1c, which relates G_1 and H_1. In this case, the vertices of both graphs are related as $\varphi_1(\mathrm{A}) = 1$, $\varphi_1(\mathrm{B}) = 2$, $\varphi_1(\mathrm{C}) = 3$, $\varphi_1(\mathrm{D}) = 4$, and $\varphi_1(\mathrm{E}) = 5$. To evaluate this solution, we consider each vertex independently. First, we evaluate A, which has two positive adjacent vertices, C and D, and the shortest paths to them are $\{1, 2, 3\}$ and $\{1, 5, 4\}$, respectively. Since A has the negative adjacent vertices B and E, assigned to host graph vertices 2 and 5, respectively, each of them generates a conflict. Therefore, vertex A has two conflicts in this embedding. Vertex D, which has a positive adjacent vertex (A) and a negative adjacent vertex (E), also generates a conflict, since E is found in the path to A. Vertices B, C, and E have no conflicts, as they have only positive or negative adjacent vertices. As a result, the objective function of this embedding is $\mathrm{CMMSA}(\varphi_1) = \max\{2, 0, 0, 1, 0\} = 2$.

The CMMSA objective function penalizes situations like the one described above and favors more balanced solutions where all elements have a similar number of conflicts. From a theoretical perspective, this new variant holds scientific interest as it transforms the CMinSA problem into a min-max situation, with specific properties and approximations, making it highly relevant for this family of problems and for any other GLP.

The main contributions of this work are twofold. First, we identify that the CMMSA problem has a solution space characterized by flat landscapes, where a large number of solutions yield identical objective function values. This property poses a significant challenge for traditional optimization techniques, which may struggle to differentiate between solutions and make meaningful progress. Second, to address this issue, we introduce a set of alternative problem formulations designed to enhance the search process. These formulations provide additional evaluation criteria that help guide the algorithm through these flat regions, enabling it to discover higher-quality solutions more effectively.

The remainder of the document is divided into four sections. Section 2 reviews the previous works that comprise the literature related to the problem and the proposed algorithms. Section 3 presents the proposed algorithmic approach in detail. This proposal is experimentally evaluated in Sect. 4 and compared with the state-of-the-art method from CMinSA adapted for this problem. Finally, Sect. 5 presents the conclusions and future work.

2   Previous Works

In this paper, we study the CMMSA problem, which is a variant of the CMinSA problem [1,13,15] which in turn is derived from the MinSA problem [7,12]. MinSA and CMinSA are closely related problems that only differ in the graph used as host. Particularly, the host graph in the case of the MinSA is a path-type graph, while in the case of the CMinSA it is a cycle-type graph.

Both MinSA and CMinSA have been studied from theoretical and practical perspectives. In particular, practical works have focused on addressing these problems using heuristic approaches.

Focusing our attention on the CMinSA problem, it was originally proposed in 2018 [1]. In that work, the authors conducted a theoretical study of the problem in which they demonstrated that, if the positive edges of the signed graph form a circular arc graph, then a solution with zero conflicts exists. Specifically, a circular arc graph is essentially an interval graph applied to a cyclic structure. This allows leveraging the circular arc graph structure to design strategies that group positive adjacent vertices in solutions, and exploit this theoretical feature in algorithmic proposals. The most recent heuristic proposal for the CMinSA was published in 2025 [15]. The proposal was based on a hyperheuristic algorithm, which, given multiple constructive methods, used *Multi-Armed Bandit* [8] to determine the best constructive method for a given instance without prior knowledge. Then, the solutions went through an improvement phase based on *Variable Neighborhood Descent* (VND) [6] with insert and swap movements. In addition, the authors also presented a new set of instances based on real-world scenarios, which complements the previous sets of synthetic instances found in the literature.

The CMMSA problem was first introduced in [14], that can be considered a preliminary work, where the authors only proposed several constructive methods for the CMMSA. The problem only differs from the CMinSA in the objective function being evaluated (the maximum number of conflicts in any vertex instead of the total sum of conflicts among all vertices).

On the other hand, it shares similar characteristics with other optimization problems such as GLP problems and min-max/max-min optimization problems. Particularly, there is a set of GLP problems where the objective is to maximize a minimum or minimize a maximum (see for instance, the *Cutwidth Minimization Problem* (CMP), the *Bandwidth Minimization Problem* (BMP), the *Anti-Bandwidth Maximization Problem* (ABMP), or the *Cyclic Cutwidth Minimization Problem* (CCMP), among others [4]).

When a solution space of an optimization problem presents a flat landscape, many solutions yield the same objective function value, making it difficult for local search methods to distinguish between them and progress effectively. To overcome this, researchers have proposed tie-breaking criteria that guide the search toward more promising regions of the solution space. In the context of GLPs, four notable works have addressed this issue. In [11], the authors introduce the VFS approach for the Cutwidth Minimization Problem, defining an `accept` function that uses multiple formulations to compare solutions during local search and neighborhood change. In [16], the main objective function is replaced with an alternative one for the Cyclic Bandwidth Sum Problem, demonstrating that optimizing a different formulation can indirectly improve the original objective. In [3], three alternative formulations are proposed for the CCMP, used exclusively in tie situations to select between equally evaluated solutions. Finally, [2] explores the correlation between three GLPs and proposes a hybrid VFS-VND approach that switches formulations upon reaching a local optimum, improving long-term search performance.

3   Algorithmic Proposal

The *Variable Formulation Search* (VFS) approach can be integrated into any variant of the VNS metaheuristic, including *Basic Variable Neighborhood Search* (BVNS), VND, *Reduced Variable Neighborhood Search*, and *General Variable Neighborhood Search* [6,11]. Its primary goal is to enhance the ability of the algorithm to navigate complex or flat regions of the solution space by leveraging multiple evaluation criteria.

In this work, we focus specifically on the integration of VFS within the BVNS framework to tackle the CMMSA. The overall structure of BVNS remains unchanged, maintaining its three main components: the *shaking* phase, the *local search*, and the *neighborhood change* mechanism. The parameter k defines the size of the neighborhood used in the shaking step. The initial solution is generated using a random constructive method.

The main innovation introduced by VFS lies in the *accept method* used during local search [11]. Instead of relying solely on the original objective function, the VFS-based version incorporates alternative formulations to guide the search when multiple solutions yield the same objective function value.

3.1   Shake Method

The shaking step in BVNS introduces diversification by randomly applying a sequence of *moves*, commonly denoted as perturbation [9]. In our case, the shake operator is performed using insertion moves. The number of insertions performed is controlled by the parameter k, which determines the magnitude of the perturbation.

An insertion move is defined as follows. Given a solution φ representing an embedding of the input vertices in the host vertices, we select one input vertex u currently assigned to host vertex $j = \varphi(u)$, and another host vertex i. The operation proceeds by: (i) Removing the embedding of u in j; (ii) reassigning each input vertex assigned to host vertices in the path of j to i in such a way that each host vertex is assigned to the next input vertex in the path; and (iii) embedding u at host vertex i.

For example, given the solution φ_1 depicted in Fig. 1c. An insertion of vertex A into host vertex 4 would yield the new solution φ_2 such that $\varphi_2(A) = 4$, $\varphi_2(B) = 1$, $\varphi_2(C) = 2$, $\varphi_2(D) = 3$, and $\varphi_2(E) = 5$.

Formally, the shaking step selects an input vertex and performs k insertion operations, generating a perturbed solution.

3.2 Local Search Method

The local search procedure explores the neighborhood of the current solution using a specific move operator until a local optimum is reached. In this case, we propose the use of a *swap* operator/move. Given a solution and two input vertices, a swap move exchanges the assignation of their corresponding host vertices in the solution.

Formally, given a solution φ, and the vertices $u, v \in V_G$ such that $\varphi(u) = i$ and $\varphi(v) = j$, a swap move results in a new solution φ' such that:

$$\varphi'(u) = j, \quad \varphi'(v) = i, \quad \varphi'(w) = \varphi(w) \; \forall w \in V_G \setminus \{u, v\}.$$

For example, given the solution φ_1 depicted in Fig. 1c a swap of the vertices A and C, embedded in the host vertices 1 and 3 respectively, would result in the solution φ_2 such that $\varphi_2(A) = 3$, $\varphi_2(B) = 2$, $\varphi_2(C) = 1$, $\varphi_2(D) = 4$, and $\varphi_2(E) = 5$.

The local search iteratively evaluates all pairwise swaps (u, v) until a better solution is found or all the possible swaps have been explored. If a new better solution is discovered, it is immediately accepted, and the local search is restarted from this improved solution until no further improvements can be made. This strategy is known in the literature as "first-improvement", and it helps the proposed method to quickly converge towards local optima.

3.3 Variable Formulation Search

The VFS schema introduced in [11] is based on BVNS. Particularly, the main modification introduced by VFS with respect to BVNS lies in the accept method used to compare solutions [11]. In a standard BVNS, during the improvement phase (i.e., the local search), a new solution is accepted if it improves the objective function; otherwise, it is rejected. In the case of VFS, it extends this process by incorporating a new `accept` method which, in some cases, evaluates the compared solutions using alternative functions. This enables the algorithm to distinguish among solutions that have the same value of the original objective function, but have a different internal structure. When many different solutions share the same value of the objective function for a specific problem, it is said that it presents a flat search landscape.

In particular, we define an `accept` method (see Algorithm 1) that replaces the standard accept method based solely on the objective function (CMMSA), which is initially used in VFS for the first comparison (lines 2–6). If the new solution (s') is better than the previous best solution (s), it is accepted (line 3); otherwise, if the quality of the new solution, based on the objective function, is worse, it is rejected (line 5). However, when both solutions yield the same objective function value (line 6), the method evaluates sequentially the compared solutions, using one by one each available function in F (lines 7–13) until one of them is able to make a distinction between the compared solutions or all the alternative functions have been tested.

Algorithm 1. Accept method VFS$_1$

1: **procedure** ACCEPT($s, s', F = [OF_1, OF_2, \ldots, OF_j]$)
2:    **if** CMMSA(s') < CMMSA(s) **then**
3:        **return true**
4:    **else if** CMMSA(s') > CMMSA(s) **then**
5:        **return false**
6:    **else**   ▷ Both solutions are tied
7:        **for** f **in** F **do**   ▷ For each alternative objective function
8:            **if** $f(s') < f(s)$ **then**
9:                **return true**   ▷ Accept immediately
10:           **else if** $f(s') > f(s)$ **then**
11:               **return false**   ▷ Reject immediately
12:           **end if**
13:       **end for**
14:       **return false**   ▷ No improvement found
15:   **end if**
16: **end procedure**

It is important to note that the order in which the alternative objective functions are evaluated is relevant. As soon as the candidate solution s' performs worse than the current solution s for any function in the list, it is immediately rejected, even if it might perform better in subsequent functions. This means that earlier functions in the list carry more influence in the decision-making process.

In this work, we denote as **VFS$_1$** the variant that uses the `accept` function as described in Algorithm 1. Additionally, we introduce a more permissive alternative denoted as **VFS$_2$**, in which the immediate rejection condition (lines 10 and 11) is removed. In this new variant, a candidate solution is accepted if it improves any of the alternative objective functions, regardless of potential degradations obtained in others. As a result, the evaluation order becomes irrelevant. In this study, we also aim to compare both variants to assess the impact of the accept method on the performance of the algorithm.

3.4   Alternative Formulations

Alternative formulations in the context of VFS were originally defined as alternative ways to evaluate a solution in which a local optimum would also be an optimum when considering the original formulation. However, when equivalent formulations are not at hand, the identification of structural properties observed in high-quality solutions and measured through mathematical functions might also help to guide search towards better solutions. Therefore, both ideas can be combined with a problem to present a flat landscape. In the case of this paper, we propose a total of four different alternative formulations for the CMMSA problem. Specifically, three of them identify structural properties, while the fourth can be considered as an alternative objective function to the CMMSA evaluation.

The first formulation (OF_1) is the one of a closely related optimization problem, the CMinSA formulation, which guides the search based on the total sum of conflicts instead of the maximum value of errors in a vertex. It is formalized as follows:

$$OF_1(\varphi) = \sum_{u \in V_G} \epsilon(u, \varphi). \tag{1}$$

As an example, we illustrate the calculation of this objective function for the solution depicted in Fig. 1c. Given the conflict values $\epsilon(A, \varphi_1) = 2$, $\epsilon(B, \varphi_1) = 0$, $\epsilon(C, \varphi_1) = 0$, $\epsilon(D, \varphi_1) = 1$, and $\epsilon(E, \varphi_1) = 0$, the evaluation of this function is the sum of conflicts at each vertex, and it is calculated as follows $OF_1(\varphi_1) = 2 + 0 + 0 + 1 + 0 = 3$. When considering this objective function, the objective is to minimize its value.

The second formulation (OF_2) is based on the total distance of the vertices to their positively connected adjacent ones. This method is based on the heuristic idea of locating positively connected vertices as close as possible to each other. This criterion can be formalized as follows:

$$OF_2(\varphi) = \sum_{u \in V_G} \sum_{v \in \Gamma^+(u)} |\psi(\varphi(u), \varphi(v))|. \tag{2}$$

We illustrate the evaluation of OF_2 with an example. Given the solution depicted in Fig. 1c, we first calculate the distance of each vertices to their positive ones. For vertex A, $\Gamma^+(A) = \{C, D\}$ leading to $|\psi(1, 3)| = 2$ and $|\psi(1, 4)| = 2$; for vertex C, $\Gamma^+(C) = \{A\}$ leading to $|\psi(3, 1)| = 2$; for vertex D, $\Gamma^+(D) = \{A\}$ leading to $|\psi(4, 1)| = 2$. Vertices B and E have no positive adjacencies. By aggregating these distances, we obtain $OF_2(\varphi_1) = 2 + 2 + 0 + 2 + 2 + 0 = 8$. As in the previous objective function, the objective is to minimize its value.

The third formulation (OF_3) is similar to OF_2, but instead of minimizing the distance with the positive adjacent vertices, it maximizes the distance with the negative ones. It can be similarly formalized as follows:

$$OF_3(\varphi) = \sum_{u \in V_G} \sum_{w \in \Gamma^-(u)} |\psi(\varphi(u), \varphi(w))|. \tag{3}$$

Again, using the solution example from Fig. 1c, we compute $OF_3(\varphi_1)$ as follows: for vertex A, the negative adjacent vertices are $\Gamma^-(A) = \{B, E\}$ leading to $|\psi(1, 2)| = 1$ and $|\psi(1, 5)| = 1$; for vertex B, $\Gamma^-(B) = \{A\}$ leading to $|\psi(2, 1)| = 1$; for vertex D, $\Gamma^-(D) = \{E\}$ leading to $|\psi(4, 5)| = 1$; for vertex E, $\Gamma^-(E) = \{A, D\}$ leading to $|\psi(5, 1)| = 1$ and $|\psi(5, 4)| = 1$. Vertex C has no negative adjacencies. By aggregating these distances, we obtain $OF_3(\varphi_1) = 1 + 1 + 1 + 0 + 1 + 1 + 1 = 6$. Let us remember that the objective is to maximize the value of this function.

Finally, the fourth alternative formulation (OF_4) is an extension of the CMMSA formulation. When two solutions share the same maximum number of conflicts, this criterion evaluates the distribution of conflicts across vertices.

Specifically, it counts how many vertices have exactly k conflicts for each possible value of k ($k \in [1, \ldots, \text{CMMSA}(\varphi)]$. A solution is considered more promising if it has fewer vertices with a higher number of conflicts. This behavior is modeled by defining a value f_k, which for a given number of conflicts k is computed as: $f_k(\varphi) = |\{u \in V_G : \epsilon(u, \varphi) = k\}|$. Using these values, the alternative formulation to be minimized is defined as:

$$OF_4(\varphi) = \sum_{k=1}^{\text{CMMSA}(\varphi)} |V|^k \cdot f_k(\varphi), \tag{4}$$

where $|V|^k$ grows exponentially with k, assigning greater weight to vertices with higher conflict counts. As a result, a small number of highly conflicted vertices penalizes the solution more than a larger number of vertices with fewer conflicts.

For the solution φ_1 depicted in Fig. 1c, the $\text{CMMSA}(\varphi_1)$ is 2. Evaluating OF_4, we find: only the vertex A has 2 conflicts, which matches the value of the $\text{CMMSA}(\varphi_1)$; then, the vertex D has 1 conflict; and the vertices B, C, E have 0 conflicts. These results in $OF_4 = 1 \cdot 5^2 + 1 \cdot 5^1 + 3 \cdot 5^0 = 33$. The objective is to minimize the value of OF_4.

It is important to note that OF_4 is the only criterion directly aligned with the original objective function, as an improvement in CMMSA implies an improvement in OF_4, and multiple improvements in OF_4 eventually lead to an improvement in CMMSA.

4   Experimental Results

This section studies the CMMSA experimentally by dealing with flat landscapes through the use of alternative formulations within the proposed VFS framework.

We selected 20 representative instances previously used in the context of SA problems, and publicly available[1]. Particularly, the data set contains four sets of input graphs: Complete, Interval, Random, and Real-world. Therefore, we have selected 5 instances per group. The number of vertices of the selected instances ranges from 70 to 190 in the Complete subset, 30 to 230 in the Interval subset, 130 to 250 in the Random subset, and 119 to 2,502 in the Real-world subset of instances. The total number of edges (positive and negative) spans from 2,415 to 17,955 in the Complete subset, 216 to 21,418 in the Interval subset, 1,677 to 15,562 in the Random subset, and 549 to 11,076 in the Real-world subset of graphs.

All experiments were conducted on a virtual machine equipped with an Intel Xeon Gold 6226R 2.90 GHz processor and 119 GB of RAM. The source code was developed in Java version 21.

The first experiment aims to analyze the structure of the neighborhood around a locally optimal solution. Starting from a randomly generated solution, we apply the local search procedure described in Sect. 3.2 until a local

---

[1] https://github.com/MRoblesR/signed-graphs.

optimum is reached. Once there, we evaluate all neighboring solutions, one by one, that could be reached with the same move operator used in the local search and record their objective function values. The distribution of these values is visualized using histograms. Specifically, Fig. 2 shows the results for four representative instances (*random_055_130x1677_20_20*, *interval_039_90x816_20_80*, *100wikipedia_adminship_election_data*, and *complete_013_30x435_100_20*), one from each subset of instances. As we can observe, in any case, a large proportion of neighboring solutions share the same objective value as the local optimum, confirming the existence of flat regions in the solution space. Specifically, for the Interval instance, 10.24% of the solutions have the same value as the current local optimum. Similarly, for the complete instance it is 10.60% and for the case of the random instance this percentage is 10.82%. Finally, for the real-world instance, we observe the largest value (29.10%) among the compared instances.

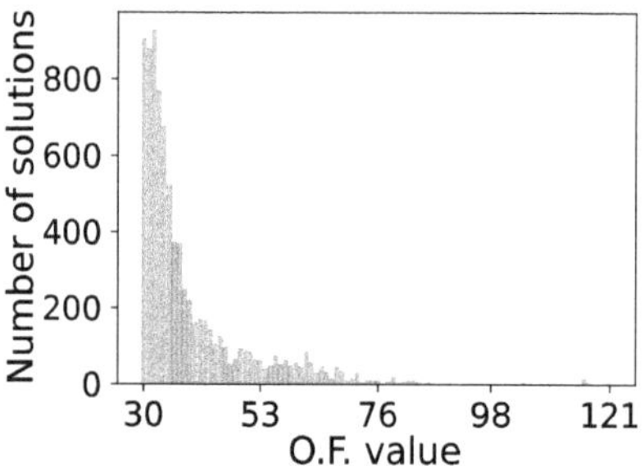

(a) *random_ 055_ 130x1677_ 20_ 20* instance. Local optimum value: 30. Best-known value for this instance = 16.

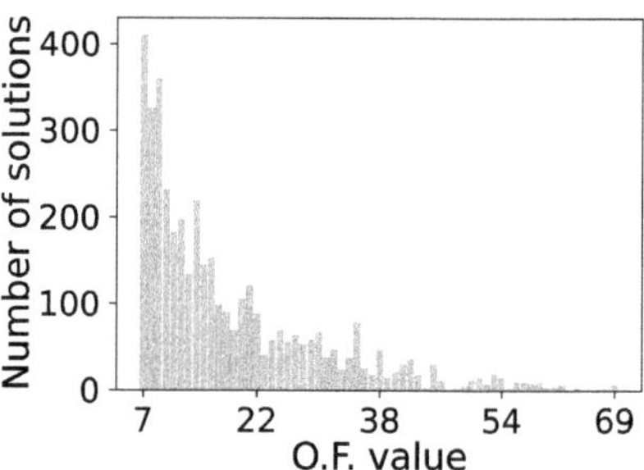

(b) *interval_ 039_ 90x816_ 20_ 80* instance. Local optimum value: 7. Best-known value for this instance = 0.

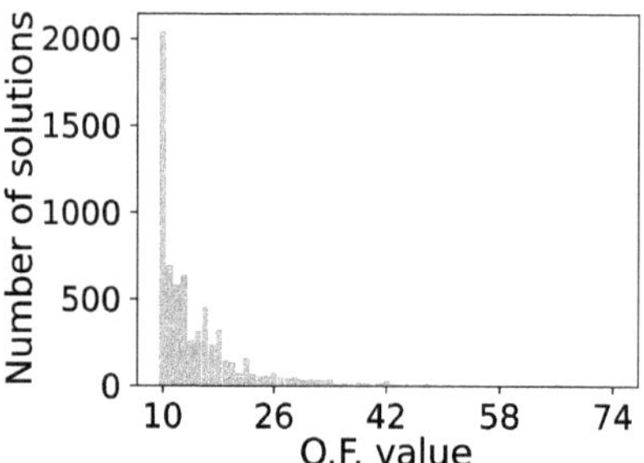

(c) *100wikipedia_ adminship_ election_ data* instance. Local optimum value: 10. Best-known value for this instance = 2.

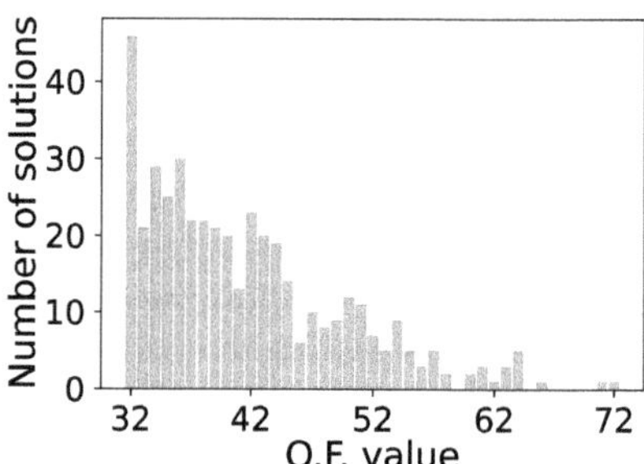

(d) *complete_ 013_ 30x435_ 100_ 20* instance. Local optimum value: 32. Best-known value for this instance = 20.

Fig. 2. Distribution of the number of solutions that share the same objective function value, considering only the neighborhood of a local optimum for each of the four representative instances considered.

Before addressing the core experiments of this study, we first tune the parameters of the VFS framework. The value of $K_{\max}$ is set to 5 based on preliminary testing with $k \in \{1, 2, \ldots, K_{max}\}$ and steps of 1 unit. For each k value, a number of insert moves δ computed as $\delta = |V_G| \cdot k \cdot 0.05$ are applied. The procedure can stop after reaching K_{max} with a time limit of one hour; if this limit is reached, the best solution found so far at that moment is returned.

Once the VFS configuration is established, we evaluate the performance of each of the five alternative formulations introduced in Sect. 3.4. As previously discussed, we consider two variants of the VFS strategy: **VFS$_1$**, which uses the strict accept method described in Algorithm 1, and **VFS$_2$**, which accepts any improvement in alternative formulations. It is worth mentioning that if only one alternative formulation is used, both variants behave identically.

Firstly, in Table 1, we compare the method that does not use any alternative formulation (i.e., BVNS) with five configurations of VFS, each of them considering only one alternative formulation simultaneously. The table reports the average objective function value (Avg. O.F.), average CPU time in seconds (CPU. T. (s)), and the number of instances (out of 20) in which each method achieved the best result (# Best).

Table 1. Performance of VFS$_1$ using one alternative formulation in addition to the base objective function.

Method	Avg. O.F.	CPU. T. (s)	# Best (20)
BVNS(CMMSA) (baseline)	520.45	**1897.60**	1
VFS: CMMSA $+ OF_1$	**472.50**	2089.70	9
VFS: CMMSA $+ OF_2$	486.20	2138.95	6
VFS: CMMSA $+ OF_3$	485.60	2156.25	2
VFS: CMMSA $+ OF_4$	478.10	2202.90	**14**

The results show that incorporating any of the alternative formulations improves the performance of the BVNS used as baseline. The most effective formulations are OF_1 (the CMinSA formulation, i.e., the sum of the errors) and OF_4 (which prioritizes reducing the number of vertices with high conflict counts). While OF_1 achieves the lowest average objective value, OF_4 obtains the highest number of best solutions. Therefore, we consider OF_4 the most robust option based on its superior consistency across instances.

Next, we extend the previous experiment by evaluating the performance of combining two alternative formulations instead of one. In this case, the distinction between VFS$_1$ and VFS$_2$ becomes relevant. For VFS$_1$, the order of the formulations matters due to the early rejection condition, resulting in 12 possible combinations. For VFS$_2$, where the rejection condition is removed, the order is irrelevant, yielding 6 unique combinations. We have developed a factorial experimental configuration, in which all the possible combinations are tested. The results are shown in the Tables 2 and 3 for VFS$_1$ and VFS$_2$, respectively.

Table 2. Results of all possible combinations using two alternative formulations for VFS_1.

Method	Avg. O.F.	CPU. T. (s)	# Best (20)
$VFS_1 : CMMSA + (OF_1 + OF_2)$	473.05	2256.95	6
$VFS_1 : CMMSA + (OF_1 + OF_3)$	472.90	2261.40	5
$VFS_1 : CMMSA + (OF_1 + OF_4)$	**472.15**	2227.45	7
$VFS_1 : CMMSA + (OF_2 + OF_1)$	483.20	2127.15	6
$VFS_1 : CMMSA + (OF_2 + OF_3)$	485.75	2117.50	6
$VFS_1 : CMMSA + (OF_2 + OF_4)$	484.35	**2018.00**	6
$VFS_1 : CMMSA + (OF_3 + OF_1)$	483.80	2305.40	2
$VFS_1 : CMMSA + (OF_3 + OF_2)$	484.00	2334.60	3
$VFS_1 : CMMSA + (OF_3 + OF_4)$	481.05	2301.05	2
$VFS_1 : CMMSA + (OF_4 + OF_1)$	480.40	2247.25	**13**
$VFS_1 : CMMSA + (OF_4 + OF_2)$	480.40	2281.45	**13**
$VFS_1 : CMMSA + (OF_4 + OF_3)$	480.85	2398.80	9

Table 3. Results of all possible combinations using two alternative formulations for VFS_2.

Method	Avg. O.F.	CPU. T. (s)	# Best (20)
$VFS_2 : CMMSA + (OF_1 + OF_2)$	477.10	2736.15	6
$VFS_2 : CMMSA + (OF_1 + OF_3)$	476.90	2906.05	5
$VFS_2 : CMMSA + (OF_1 + OF_4)$	**468.55**	**2551.30**	**15**
$VFS_2 : CMMSA + (OF_2 + OF_3)$	476.20	2911.05	6
$VFS_2 : CMMSA + (OF_2 + OF_4)$	474.50	2556.55	7
$VFS_2 : CMMSA + (OF_3 + OF_4)$	472.45	2911.45	9

In the case of VFS_1, the results show that combining two formulations can further improve performance in some cases. The best average objective value is achieved by $OF_1 + OF_4$, while the highest number of best solutions is obtained by $OF_4 + OF_1$ and $OF_4 + OF_2$. This confirms the strong individual performance of OF_1 and OF_4, and suggests that their combination is particularly effective. However, the enhancement in terms of quality is very little in comparison to the extra needed time.

In the case of VFS_2, again, the method with the alternative formulations OF_1 and OF_4 is the one that obtains the best results, in a similar fashion to the results obtained with VFS_1. In terms of CPU time, as a consequence of removing the immediate rejection condition, the VFS_2 requires more time than VFS_1. Specifically, this happens because i) VFS_2 generally needs to evaluate more alternative formulations and ii) the possibility of worsening an alternative formulation, and later improving it, might produce looping in the local search if both formulations are opposed.

Finally, in Table 4, we compare the best-performing configurations, including the baseline BVNS algorithm that does not use alternative formulations. Note that for this comparison, in contrast to previous experiments, the column of # Best refers to the number of best-known solutions obtained with respect to any algorithm in the paper, instead of just the ones in this experiment.

Table 4. Comparison of the best-performing algorithms.

Method	Avg. O.F.	CPU. T. (s)	# Best
BVNS(CMMSA) (baseline)	520.45	**1897.60**	1
VFS: $CMMSA + OF_1$	472.50	2089.70	6
VFS: $CMMSA + OF_4$	478.10	2202.90	**10**
VFS_1: $CMMSA + (OF_1 + OF_4)$	472.15	2227.45	6
VFS_2: $CMMSA + (OF_1 + OF_4)$	**468.55**	2551.30	**10**

As it can observed, all VFS-based methods outperform the baseline VFS_1 with OF_4 achieve the highest number of best solutions found, which is also matched by VFS_2 with $OF_1 + OF_4$. In this case, this variant also achieves the best average objective value. Overall, VFS_2 with $CMMSA + (OF_1 + OF_4)$ is the method in terms of objective function and number of best solutions found; in contrast, it is the slowest method among all five.

To assert whether there are statistically significant differences, we employ the Wilcoxon signed-rank test applied by pairs. As a result, all configurations of VFS tested not only outperform the BVNS but also achieve statistically significant differences with respect to it. However, we do not find statistically significant differences among the four proposed VFS variants. Therefore, the use of VFS is recommended for this problem, and any of the best configurations found is suitable to tackle the problem.

5   Conclusions and Future Work

In this work, we addressed the CMMSA problem, a graph layout problem characterized by flat landscapes in the solution space, where many solutions share the same objective function value. To improve search effectiveness in this context, we proposed a VNS-based approach based on the VFS variant of the methodology, which incorporates the use of alternative formulations to guide the search for solutions for an optimization problem.

We evaluated four alternative formulations for the CMMSA and found that the use of an alternative formulation within a VFS schema, outperforms the BVNS, which only considers the original formulation of the problem. Among the tested formulations, the most effective are those that minimize the total number of conflicts (OF_1) and those that reduce the number of vertices with the maximum number of conflicts (OF_4). These criteria have been shown to help the algorithm escape flat regions and identify promising solutions during the search. We also compared the use of a single alternative formulation versus the combination of two or more. The results showed that the use of one well-chosen formulation is sufficient, as combining two does not consistently improve performance.

In future work, we propose exploring learning-based strategies (i.e., machine learning or reinforcement learning) to dynamically select the most appropriate formulation during the search for each particular instance.

Acknowledgments. This research has been partially supported by grants with ref. ID: CIRMA-CM Ref. TEC-2024/COM-404; PID2024-160226OB-C22; PID2021-125709OA-C22; TSI-100930-2023-3 (MCA07), and MCA06.

References

1. Benítez, F., Aracena, J., Caro, C.T.: The sitting closer to friends than enemies problem in the circumference (2018). https://doi.org/10.48550/arXiv.1811.02699
2. Cavero, S., Colmenar, J.M., Pardo, E.G.: A variable formulation search approach for three graph layout problems. In: Sevaux, M., Olteanu, A.L., Pardo, E.G., Sifaleras, A., Makboul, S. (eds.) MIC 2024. LNCS, vol. 14753, pp. 390–396. Springer, Cham (2024). https://doi.org/10.1007/978-3-031-62912-9_38
3. Cavero, S., Pardo, E.G., Duarte, A.: Influence of the alternative objective functions in the optimization of the cyclic cutwidth minimization problem. In: Alba, E., et al. (eds.) CAEPIA 2021. LNCS (LNAI), vol. 12882, pp. 139–149. Springer, Cham (2021). https://doi.org/10.1007/978-3-030-85713-4_14
4. Cavero, S., Pardo, E.G., Duarte, A., Martí, R.: Graph Layout Problems, pp. 1–31. Springer, Cham (2025). https://doi.org/10.1007/978-3-319-07153-4_45-2
5. Díaz, J., Petit, J., Serna, M.: A survey of graph layout problems. ACM Comput. Surv. **34**(3), 313–356 (2002). https://doi.org/10.1145/568522.568523
6. Hansen, P., Mladenović, N., Brimberg, J., Pérez, J.A.M.: Variable neighborhood search. In: Gendreau, M., Potvin, J.-Y. (eds.) Handbook of Metaheuristics. ISORMS, vol. 272, pp. 57–97. Springer, Cham (2019). https://doi.org/10.1007/978-3-319-91086-4_3
7. Kermarrec, A.-M., Thraves, C.: Can everybody sit closer to their friends than their enemies? In: Murlak, F., Sankowski, P. (eds.) MFCS 2011. LNCS, vol. 6907, pp. 388–399. Springer, Heidelberg (2011). https://doi.org/10.1007/978-3-642-22993-0_36
8. Meidani, K., Mirjalili, S., Barati Farimani, A.: MAB-OS: multi-armed bandits metaheuristic optimizer selection. Appl. Soft Comput. **128**, 109452 (2022). https://doi.org/10.1016/j.asoc.2022.109452
9. Mladenović, N., Urošević, D., Hanafi, S.: Variable neighborhood search for the travelling deliveryman problem. 4OR **11**(1), 57–73 (2013). https://doi.org/10.1007/s10288-012-0212-1
10. Newton, A.R.: Computer-aided design of VLSI circuits. Proc. IEEE **69**(10), 1189–1199 (1981)

11. Pardo, E.G., Mladenović, N., Pantrigo, J.J., Duarte, A.: Variable Formulation Search for the Cutwidth Minimization Problem. Appl. Soft Comput. **13**(5), 2242–2252 (2013). https://doi.org/10.1016/j.asoc.2013.01.016

12. Pardo, E.G., Soto, M., Thraves, C.: Embedding signed graphs in the line: Heuristics to solve MinSA problem. J. Comb. Optim. **29**(2), 451–471 (2015). https://doi.org/10.1007/s10878-013-9604-1

13. Robles, M., Cavero, S., Pardo, E.G.: BVNS for the minimum sitting arrangement problem in a cycle. In: Sleptchenko, A., Sifaleras, A., Hansen, P. (eds.) ICVNS 2022. LNCS, vol. 13863, pp. 82–96. Springer, Cham (2023). https://doi.org/10.1007/978-3-031-34500-5_7

14. Robles, M., Cavero, S., Pardo, E.G.: Heurísticas multiarranque para el problema de cmmsa. In: XVI Congreso Español de Metaheurísticas, Algoritmos Evolutivos y Bioinspirados (2025)

15. Robles, M., Cavero, S., Pardo, E.G., Cordón, O.: Multi-armed bandit for the cyclic minimum sitting arrangement problem. Comput. Oper. Res. **179**, 107034 (2025). https://doi.org/10.1016/j.cor.2025.107034

16. Rodriguez-Tello, E., Lardeux, F., Duarte, A., Narvaez-Teran, V.: Alternative evaluation functions for the cyclic bandwidth sum problem. Eur. J. Oper. Res. **273**(3), 904–919 (2019). https://doi.org/10.1016/j.ejor.2018.09.031

17. Zaslavsky, T.: A mathematical bibliography of signed and gain graphs and allied areas. Electron. J. Comb. DS8 (2018). https://doi.org/10.37236/29

Large Language Models for Metaheuristic Implementation: A Case Study with Variable Neighborhood Search

Aidan Riordan[1], Daniel Aloise[2(✉)], and Raúl Martín-Santamaría[3]

[1] Northeastern University, Boston, USA
`riordan.a@northeastern.edu`
[2] GERAD and Polytechnique Montréal, Montreal, Canada
`daniel.aloise@polymtl.ca`
[3] Universidad Rey Juan Carlos, Móstoles, Spain
`raul.martin@urjc.es`

Abstract. Metaheuristics are widely used in combinatorial optimization to address large-scale problems but often require considerable expertise and engineering effort. In this paper, we investigate whether recent large language models (LLMs) can translate state-of-the-art algorithmic descriptions into performant code with minimal human intervention. We introduce a generalizable prompting framework: Context, Communication, and Iteration (CCI), and apply it to replicate a Variable Neighborhood Search for the balanced minimum sum-of-squares clustering problem using Gemini 2.5 Pro and Claude Sonnet 4. According to our computational experiments, Gemini achieved solutions with a mean difference of 0.032% from the reference implementation (median: 0.009%), being statistically non-inferior within a 0.01% margin. Notably, Gemini corrected a missing mathematical term in the neighborhood evaluation formula described in the source paper, whereas Claude did not. Our findings suggest that LLM-assisted engineering is a viable pathway to facilitate algorithm replication, making high-performance methods more accessible to researchers and allowing them to tackle larger and more complex problems.

Keywords: Combinatorial Optimization · Metaheuristics · Large Language Models · Variable Neighborhood Search · Code Generation

1 Introduction

Metaheuristics are high-level frameworks used to find near-optimal solutions for complex combinatorial optimization problems where exact methods are computationally prohibitive [1–3]. However, implementing a state-of-the-art metaheuristic is sometimes a significant undertaking, requiring deep domain knowledge and specialized programming skills to tailor algorithms and data structures for efficient exploration of the solution space.

Large Language Models (LLMs) have emerged as powerful AI systems capable of generating human-like text and code, transforming domains from content creation to software development. Although their pattern-based reasoning constrains their ability to solve optimization problems directly, their strengths in code generation and language understanding offer an opportunity to streamline the algorithm development pipeline [4]. To overcome issues related to reliability or safety, LLMs can incorporate human feedback or control [5] or integrate evaluation-driven approaches [6].

Recent studies have begun exploring the integration of LLMs into metaheuristic algorithms. Some approaches embed LLMs as search operators within evolutionary methods [7], use them to guide the search process by identifying instance-specific patterns [8], or employ them to generate problem-specific heuristics online [9]. Others have used LLMs as zero-shot surrogate models to guide solution selection [10] or combined them with evolutionary strategies to automatically refine heuristics [11–14].

While these approaches demonstrate LLMs' potential to enhance metaheuristic performance, they leave unaddressed the challenge of replicating published algorithms whose implementation details are often ambiguous or incomplete. Algorithmic replication is essential for reproducibility and building upon prior work, yet remains a barrier for researchers without deep domain expertise.

In this work, we investigate a different role for LLMs: as a tool for replicating a complete, state-of-the-art metaheuristic directly from a scientific paper. Our central research question is whether an off-the-shelf LLM can translate a published algorithm description into functional, performant code with minimal, non-expert human guidance. As a case study, we task LLMs with implementing a Variable Neighborhood Search (VNS) [15] for the balanced minimum sum-of-Squares clustering (BMSSC) problem [16].

Our main contributions are:

1. A generalizable framework – Context, Communication, and Iteration (CCI) – for guiding an LLM through a complex code generation task.
2. A demonstration that a modern LLM can successfully replicate a non-trivial metaheuristic, achieving performance statistically comparable to the original implementation.
3. An analysis revealing how different LLMs handle incomplete specifications, through literal translation versus pattern-based inference, affecting replication fidelity.

The remainder of this paper is organized as follows. Section 2 presents our proposed prompting methodology. Section 3 applies the methodology through a VNS case study for BMSSC. Section 4 compares performance results. Section 5 concludes with key findings and directions for future work.

2   Methodology

Our LLM-assisted coding methodology is designed as an iterative collaboration between a human user and an LLM, where the user provides high-level guidance and debugging feedback, and the LLM handles the detailed implementation. First, the main concepts of the methodology are introduced, followed by an in-depth explanation of each component.

2.1 Main Concepts

The methodology builds on our proposed Context, Communication, and Iteration (CCI) framework, which guides an LLM through complex code generation tasks.

Context. The LLM needs to be provided with all relevant information related to the given task. In our case, the context includes the source research paper describing the problem objective and its constraints, the algorithm description, a basic project structure, and supporting code files. The supporting code files should define how to parse the instance file and function interfaces or signatures to facilitate integration, but should not contain any algorithm implementation. Specific care must be taken to avoid guiding the LLM by including data structures specific to certain algorithms. For example, defining a candidate list might guide the LLM to implement a GRASP like approach.

Communication. Instructions must be provided to the LLM describing the task to perform. Because the general task is too complex to be performed in a single step, the overall task is decomposed into smaller, manageable steps. In each of them, the user provides a clear, specific prompt defining the desired outcome for the current step. This allows testing each step independently during development.

Iteration. The development process is cyclic. At the start of each cycle, the user compiles and runs the code generated by the LLM, returning any error messages or performance metrics back to the LLM. This feedback loop allows the model to diagnose issues and refine its implementation.

2.2 Workflow

A summary of our methodology workflow is represented by Fig. 1. First, the data that is going to be provided as context needs to be prepared. We need to provide the problem definition, including its objective function and constraints. This is accomplished by providing the LLM with the full text from the original research paper, removing only the bibliography section.

A project skeleton is also provided as context. This project skeleton contains basic method definitions (for example, a constructive method receives an instance and returns a solution) and a minimal instance and solution structure that the LLM can modify. No algorithm component implementation is provided to the LLM.

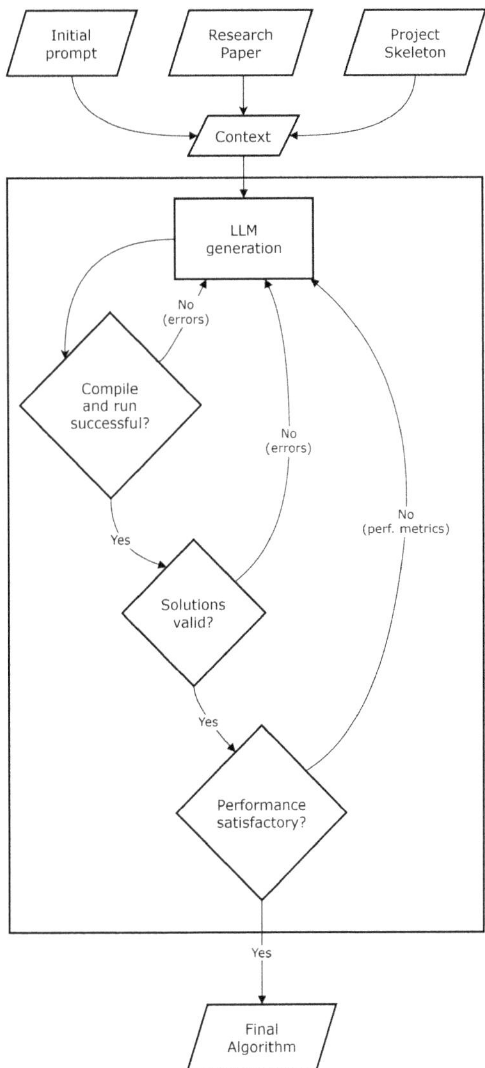

Fig. 1. Iterative methodology for LLM-assisted coding. The user provides context and a high-level goal, and then iterates with the LLM providing feedback on the generated code until a satisfactory implementation is achieved.

In the first iteration, the LLM is provided with pre-written prompts including the research paper and the supporting project skeleton as context. The LLM will then generate an initial implementation, which is copied into the development environment. In addition to the files generated by the LLM, the project contains supporting files which implement the necessary functionality to load instances from files, validate and export solutions, and generate a report with different execution metrics. Example metrics in the context of metaheuristics include execution times, average gap to the best known value, and the number of times the search method reaches the best known value.

Then, the iterative process begins. In each iteration, first, the project is compiled. If compilation is not successful, the compilation error is passed to the LLM, and a fix is requested, restarting the process. If compilation is successful, the generated algorithm is executed. If execution fails (for example, the process might crash due to invalid memory accesses), information about the crash is gathered and provided to the LLM requesting a fix. During the execution, metrics are collected for each component, in order to analyze the generated algorithm performance.

Once the experiment finishes successfully, all generated solutions are validated, checking that they meet all the problem constraints, and that the reported objective function is correct. This validator acts as a safeguard against both misinterpretations and hallucinations, and therefore must not be implemented by the LLM. Any mismatch detected in this step is reported to the LLM, detailing why the solution is not correct.

If the execution terminates successfully, and solutions are correct, a summary of the results will be given to the LLM in the next prompt. The LLM can evolve any part of the algorithm proposal, and the process is then repeated until the obtained results are satisfactory.

3   Case Study: A VNS for BMSSC

In this section, we present a case study of our CCI framework for metaheuristic implementation. In particular, we target the coding of a VNS, including its corresponding local search and perturbation components.

3.1   Problem and Algorithm Selection

In order to assess our methodology, we have selected BMSSC as the testbed optimization problem for our approach. Given a set $P = \{p_1, \ldots, p_n\}$ of data points in Euclidean space, the goal of BMSSC is to partition these points into k balanced clusters such that the sum of squared Euclidean distances between each point and the centroid of its assigned cluster is minimized. The balance constraint means that all clusters must have the same number of points assigned to them, unless $n \mod k \neq 0$, in which case there will be $n \mod k$ clusters of size $\lceil n/k \rceil$, and $k - (n \mod k)$ clusters of size $\lfloor n/k \rfloor$. The BMSSC is known to be NP-hard for $n/k \geq 3$ in general dimension [17]. Figure 2 shows two clustering examples for the same dataset. In Fig. 2a, the optimal point assignment is shown, while on the right Fig. 2b shows a suboptimal assignment, in which points have a greater average distance to the centroid of the cluster they are assigned to.

Two state-of-the-art metaheuristic for the BMSSC have been developed, a strategic oscillation approach combined with GRASP, developed in Java [18];

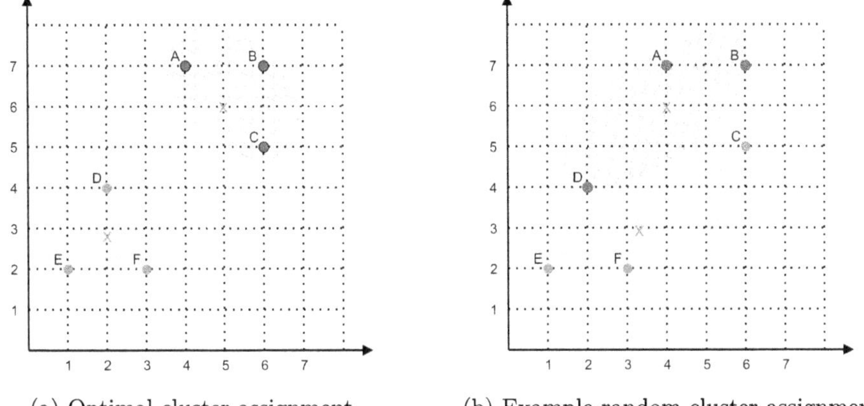

(a) Optimal cluster assignment   (b) Example random cluster assignment

Fig. 2. Two clustering examples for the same points dataset. Each point has the color of the cluster it is assigned to, whose area is delimited by an oval. A red "X' (Color figure online)' represents the centroid of each cluster.

and a VNS developed in C++, following the "less-is-more" (LIMA) design principles for heuristics [16], proposed in [19], which advocate using a simple single neighborhood structure for search space exploration. In that work, the single neighborhood used for a solution consists of all solutions obtained by swapping pairs of data points between clusters. Given its effectiveness and simplicity, we selected this VNS heuristic as the comparison baseline for the LLM-generated VNS produced by our CCI framework.

Another key reason for using the VNS developed in [16] as a testbed for our experiments is that the authors inadvertently omitted a term when computing the cost variation resulting from swapping a pair of data points.

The objective function of the BMSSC clustering problem can be decomposed into two terms, since the cardinality of the clusters are known a priori. It can be expressed as follows:

$$\frac{1}{\left\lceil \frac{n}{k} \right\rceil} \sum_{j \in \mu} \sum_{i=1}^{n-1} \sum_{\ell=i+1}^{n} \|p_i - p_\ell\|^2 w_{ij} w_{\ell j} + \frac{1}{\left\lfloor \frac{n}{k} \right\rfloor} \sum_{j \in \eta} \sum_{i=1}^{n-1} \sum_{\ell=i+1}^{n} \|p_i - p_\ell\|^2 w_{ij} w_{\ell j}, \quad (1)$$

where μ and η are the index set of clusters whose cardinality is $\left\lceil \frac{n}{k} \right\rceil$ and $\left\lfloor \frac{n}{k} \right\rfloor$, respectively, and the decision variables w_{ij} are used to express the assignment of the data point p_i to cluster j.

In [16], the authors observe that the cost of a cluster j^* is a quadratic term, given by $w_{j^*} Q w_{j^*}^T$, divided by the cardinality of cluster j^* (a constant), where $w_{j^*} = (w_{1j^*}, \ldots, w_{nj^*})$ and $Q = (q_{ab})$ with $q_{ab} = \|p_a - p_b\|^2/2$. Given that, they provide a formula for computing in constant time the cost variation of adding a point $p_\ell \in P$ to cluster j^*, which is expressed as:

$$\Delta_{\ell j^*} = \frac{2}{size(j^*)} \sum_{i=1}^{n} q_{\ell i} w_{ij^*}, \tag{2}$$

where $size(j^*)$ denotes the cardinality of cluster j^*.

Likewise, the authors provide a closed formula for computing the cost variation of removing $p_m \in P$ from cluster j^*, being expressed as:

$$\Delta_{mj^*} = -\frac{2}{size(j^*)} \sum_{i=1}^{n} q_{mi} w_{ij^*}. \tag{3}$$

In the sequel, the authors declare:

> *"Thus, the variation in the cost of cluster j^*, from which a point p_m is removed and a point p_ℓ is added, is calculated by a closed expression given by $\Delta_{\ell j^*} + \Delta_{mj^*}$."*

A closer look into that statement reveals the existence of a missing term for correctly computing the cost variation when adding a point p_ℓ from cluster j^* while removing point p_m. The missing term is $-\frac{2}{size(j^*)} q_{\ell m}$, which accounts for the direct pairwise contribution between p_ℓ and p_m that arises when one replaces the other in the cluster j^* (remark that w_{j^*} is not updated while evaluating the cost of the move). Nevertheless, the results reported in [16] are still correct, since the authors took into account that term in their C++ VNS implementation.

This setting provides an ideal testbed for our experiments with LLMs within the CCI framework, aimed to replicate metaheuristic algorithms. In particular, we investigate whether the LLMs are able to identify the missing term and thereby produce a correct implementation of the VNS heuristic described in [16].

3.2 Experimental Setup and Initial Prompts

We conducted the experiment with both Anthropic's Claude Sonnet 4 (extended thinking mode) and Gemini 2.5 Pro. We present Claude's development process in detail below as it illustrates the iterative debugging workflow, though Gemini's superior performance make it our primary focus.

All chats were configured to not retain or memorize information in each session to avoid contaminating the results with knowledge from different experiment runs. The interaction followed a structured decomposition of the problem. First, a prompt (Fig. 3) clearly defined the objective of developing the VNS algorithm as described by the paper and tasked the LLM with summarizing the research paper, which was attached as a PDF file.

Next, our second prompt (Fig. 4) provided the essential implementation files (Solution.h, Solution.cpp, DistanceMatrix.h, DistanceMatrix.cpp, Pair.h) and the header files for the modules to be implemented (Vns.h, LocalSearch.h). These header files provide a blueprint for the functions required in the files to be implemented. We then asked the LLM to describe how these components would interact.

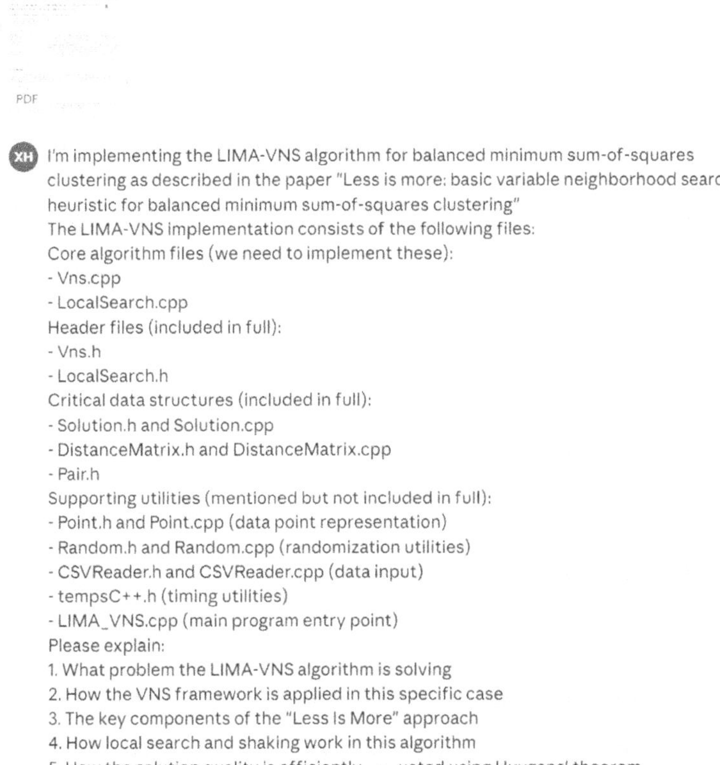

Fig. 3. Initial prompt to establish role and problem.

Then we isolated the most complex part of the algorithm, i.e., the exploration of the swap neighborhood, which is the core part of the local search procedure. We asked the LLM to first generate this procedure in pseudocode before requesting the complete C++ implementation.

3.3   Iterative Development and Error Resolution

After the initial code was generated, the process moved to an iterative debugging workflow, predefined by the researcher (Fig. 5). This phase mirrored a traditional software development cycle of implementation, testing, and refinement. In total, the experiment for Claude took nine back-and-forth interactions to produce its implementation. The errors we resolved fell into two clear categories: (i) compilation and syntax error, and (ii) logical and runtime errors. Each category highlights different aspects of the LLM's capabilities and limitations.

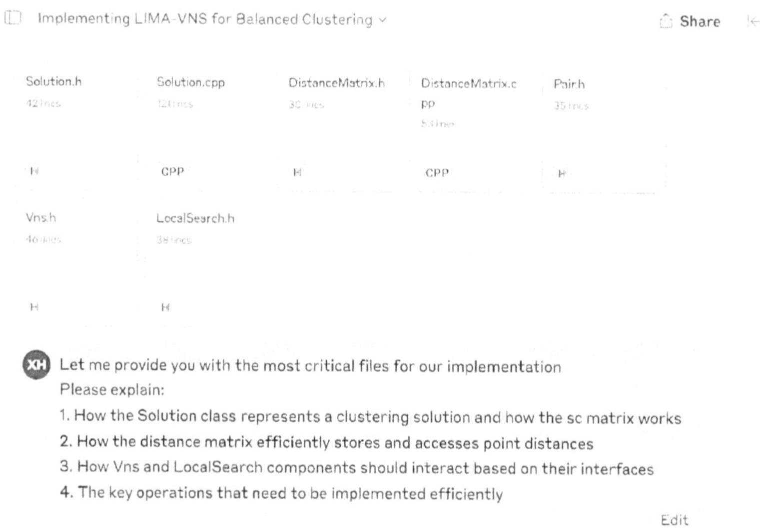

Fig. 4. Prompt providing code context and requesting an implementation plan.

After I compile and run the code, I may encounter errors. Here's
the process we'll follow:

1. *I'll compile the code first to check for syntax or type errors*
2. *If compilation succeeds, I'll run tests with the same datasets used in the paper*
3. *If I encounter any errors, I'll share the exact error messages*
4. *We'll iteratively fix issues until the implementation runs correctly*

What are some common issues or edge cases I should watch for when testing this implementation?

Fig. 5. Prompt—Establishing the Iterative Debugging Workflow.

3.3.1   Resolving Compilation and Syntax Errors

The first challenges involved classic compilation errors. This initial stage of resolving simple syntax issues was a necessary step common to both the Claude and Gemini experiments. The initial compilation failed due to API hallucinations, where the LLM used statistically common but incorrect method names (e.g., `getElapsedTime()`, `randp(int)`) instead of adhering to the function signatures provided in the project's header files. Providing the verbatim compiler error, which included a helpful "`did you mean GetTime?`" suggestion, allowed the LLM to immediately correct the function call.

This cycle also revealed the LLM's lack of proactive generalization. After the LLM fixed the API errors in `LocalSearch.cpp`, the model did not apply the correction to `Vns.cpp`, which required a separate, identical debugging iteration. The final compilation issue was a more human-like mistake corresponding to a simple variable scope error that was also automatically fixed by the LLM by analyzing the error message generated by the compiler.

3.3.2 Debugging Logical and Runtime Errors

Once the code compiled successfully, testing revealed deeper logical flaws. The first working version produced mathematically impossible negative objective function values for BMSSC solutions. The root cause was a missing term in the cost-variation formula of a swap move, as detailed in Sect. 3.1. The LLM had transcribed this formula exactly as presented in [16], where the author inadvertently omitted the necessary correction term. This omission not only caused the negative objective values but also led to the gradual drift of solution values observed in later iterations.

After the LLM identified and corrected this flaw, the next version passed the initial checks, but eventually failed after many iterations due to a residual numerical stability issue related to floating-point rounding errors. This problem was diagnosed by the LLM after analyzing the objective function evolution logs, and seeing that the incrementally calculated solution value slowly diverged from the recalculated one. In response, the LLM implemented a hybrid strategy, maintaining efficient incremental updates during local search and executing a full recalculation at regular intervals corresponding to a fixed number of VNS iterations to correct accumulated numerical errors.

Next, an integration bug in the output reporting was addressed. While the internal timer logged correct elapsed times during execution, the final summary always reported 0.00 s. Without seeing the main program's code, the LLM inferred a data communication issue between its module and the unseen calling function. It resolved this by explicitly storing the total execution time in the shared `bestSolution` object, making the data accessible for the final report.

The subsequent implementation ran successfully on 156 of 160 test runs, failing on four with a heap-buffer-overflow error, caused by an invalid memory access. This memory access violation was reliably caught using `AddressSanitizer` [20]. The root cause was a classic "off-by-one" error in the randomization logic, triggered when the `randp()` function returned an upper bound of exactly 1.0. The LLM fixed this by adding a simple safety check to clamp the index, avoiding this way the invalid memory access.

It is important to note that the way Claude worked around the missing term in the cost-variation formula of a swap move, i.e. by recalculating the cost of the current solution periodically after every fixed number of VNS iterations, does not constitute a *correct* local descent. Indeed, ignoring the missing term can cause the algorithm to perform swap moves that it incorrectly perceives as improving, even when they are not. This experience highlights a critical challenge in LLM-assisted replication: a model may produce a perfectly literal—yet functionally incorrect—implementation when the source description is ambiguous or incomplete.

In contrast, Gemini correctly incorporated the interaction term into its algorithm without specification. An analysis of the model's generation process suggests this output resulted from a procedure that synthesized the formula from the paper's description. The model's "thinking" log, a trace of its generation process prior to finalizing a response, shows a prioritization of these mechanics over the explicit formula, stating, *"I'm carefully reviewing the distance corrections when a point moves clusters"*. This procedural focus on modeling the fundamental point exchange, rather than transcribing the paper's simplified abstraction, produced the correct and complete formula, thereby avoiding the omission in the source material. Furthermore, when queried post-implementation about its formula derivation, Gemini provided a detailed mathematical justification for including the interaction term. Notably, it explained the term as a necessary correction *"we must subtract the distance from p_ℓ to the now-removed point p_m"*, while simultaneously claiming this matched the paper's specification. This shows that not only did its generation process produce a correct implementation, but it could also articulate the formal reasoning behind it, even while being unaware that it had corrected the omission in [16].

4   Performance Analysis

We compared the final LLM-generated VNS implementations from Gemini (denoted LLM-VNS) against the original authors' VNS code [16] (denoted Orig-VNS) on 16 benchmark datasets from the UCI Machine Learning Repository, presented in Table 1. We did not use Claude's implementation of the VNS of [16] given the error in its local search procedure explained in Sect. 3.3.2.

The main difference between LLM-VNS and Orig-VNS comes from another unspecified detail in the paper. The original code when performing t swaps uses an auxiliary array to ensure that the data points involved in the swaps are all unique. Gemini's code performs t random swaps, ensuring the two points in any single swap are different and in different clusters, but does not prevent the same point from being chosen in a subsequent swap within the same shake. To ensure a fair comparison, both algorithms were executed 10 times on each dataset, starting from the same random initial solutions, and using as stopping criterion the time limits defined in [16].

Figure 6 illustrates the distribution of the paired relative difference (%) in the final solution value obtained by Gemini's LLM-VNS and the final solution obtained by Orig-VNS across the 160 paired executions. The mean and median relative differences were low, 0.032% and 0.009%, respectively, with a standard deviation of 0.053%, confirming that in at least half of all test runs, the Gemini-generated code achieved a final solution value identical to that of the original implementation. Analyzing execution times, the average execution time difference between the original implementation and the LLM is 21.1%, with a median value of 6.1%.

Table 1. List of used datasets from the UCI Machine Learning repository [21]

Dataset	n	s	k
Iris	150	4	3
Wine	178	13	3
Glass	214	10	7
Thyroid	215	5	3
Ionosphere	351	34	2
Libra	360	90	15
User knowledge	403	5	4
Body measurements	507	5	2
Water treatment plant	527	38	13
Breast cancer	569	30	2
Synthetic control	600	60	6
Vehicle	846	18	6
Vowel recognition	990	10	11
Yeast	1484	8	10
Multiple features	2000	240	7
Image segmentation	2310	19	7

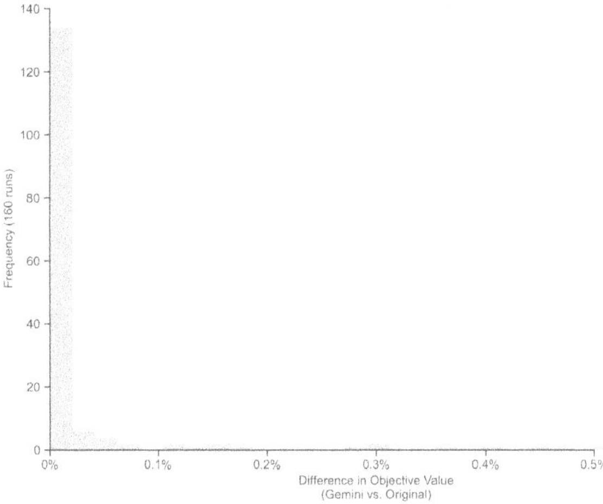

Fig. 6. Histogram of paired relative differences (in %) between solutions obtained by Gemini's LLM-VNS and Orig-VNS.

Table 2 presents the average relative differences between Gemini and the Original implementation for each instance. Most differences are negligible. The largest discrepancy occurred in the Image segmentation and Multiple `features` datasets, where the solutions produced by LLM-VNS were, on average, 0.203% and 0.116% higher than those of Orig-VNS, respectively.

Table 2. Paired relative differences (in %) between solutions obtained by Gemini's LLM-VNS and Orig-VNS. A positive difference indicates that the original algorithm outperformed the LLM version.

Dataset	Diff (%)
Iris	0.000
Wine	+0.004
Glass	+0.010
Thyroid	+0.058
Ionosphere	0.000
Libra	+0.046
User knowledge	+0.023
Body measurements	0.000
Water treatment plant	+0.017
Breast cancer	+0.029
Synthetic control	+0.008
Vehicle	0.000
Vowel recognition	0.000
Yeast	0.000
Multiple features	+0.116
Image segmentation	+0.203

To compare the fidelity of Gemini's replicated algorithm, we performed a one-sided Wilcoxon signed-rank test for non-inferiority on the complete set of paired runs. We used a non-parametric test because the data was not normally distributed as confirmed by a ShapiroWilk test ($W = 0.792, p = 0.002$).

The paired percent difference is defined as $d_i = 100 \times \frac{f_{\text{LLM-VNS}} - f_{\text{Orig-VNS}}}{f_{\text{Orig-VNS}}}$, and we set a strict non-inferiority margin of $\delta = 0.01\%$ for the signed-rank test. Thus, the evaluated hypotheses were:

- H_0 (Null Hypothesis): The median performance of the LLM code is worse than the reference by at least 0.01%.
- H_1 (Alternative Hypothesis): The LLM code is non-inferior within the 0.01% margin.

For the Gemini-generated algorithm, the test produced a p-value of 2×10^{-6}, leading us to reject the null hypothesis. This provides statistically significant evidence that the Gemini-VNS code is non-inferior to the original implementation within this strict performance margin.

5   Conclusions and Future Work

This work successfully demonstrates that large language models, guided by a structured human-in-the-loop process, can replicate a metaheuristic from a research paper to a level of performance statistically comparable to the original implementation. Our iterative framework allowed the LLM to resolve its errors and refine its implementation. This approach has the potential to democratize access to high-performance optimization algorithms, enhance reproducibility, and accelerate research by lowering the barrier to replicating and building upon prior work.

Our analysis provides valuable insights for human-AI collaboration. The improved performance of Gemini 2.5 Pro over Claude Sonnet 4.0 suggests that as LLMs improve, their capability for complex algorithmic coding will also improve. However, our experience with Claude highlights a critical risk: missing or ambiguous implementation details in source materials can lead LLMs to produce theoretically flawed replications that appear plausible but contain unforeseen errors. The key differentiator between the different LLMs performance was the capacity to correct implicit context missing from the provided source.

Future work will aim to test the LLM's ability to design and implement novel metaheuristics with minimal human intervention, potentially using agentic frameworks like AlphaEvolve [22] for discovery. We are also working to automate the iterative validation process by developing a Model Context Protocol (MCP)[1] module in the Mork optimization framework[2] to allow LLMs to run experiments and gather feedback autonomously. Moreover, we will extend the analysis to additional problem families and different metaheuristics, to further demonstrate the robustness of the CCI approach.

Supplementary information

Code Availability. The complete implementation of the LLM-generated metaheuristic, which replicates the LIMA-VNS for the balanced minimum sum-of-squares clustering (BMSSC) problem, is available at our GitHub repository.

Experiment Artifacts and Datasets All experiment artifacts, including interactive chat logs demonstrating the step-by-step process of LLM-assisted code development, instance files, and generated code are archived in the Zenodo open repository, with DOI 10.5281/zenodo.17981976.

Acknowledgements. This work has been partially supported by the Spanish Ministerio de Ciencia e Innovación (MCIN/AEI/10.13039/501100011033) under grant refs. RED2022-134480-T, PID2021-125709OA-C22 and by ERDF A way of making Europe; and Comunidad Autónoma de Madrid with grant ref. TEC-2024/COM-404; by the Natural Sciences and Engineering Research Council of Canada (NSERC), grant 2023-04466; and by the National Science Foundation (NSF) Research Traineeship (NRT) Program under grant 2244340.

---

[1] https://github.com/modelcontextprotocol.
[2] https://github.com/mork-optimization/mork.

References

1. Glover, F.: Tabu search: a tutorial. Interfaces **20**(4), 74–94 (1990)
2. Holland, J.H.: Genetic algorithms. Sci. Am. **267**(1), 66–73 (1992)
3. Resende, M.G., Ribeiro, C.C.: Optimization by GRASP. Springer, New York (2016). https://doi.org/10.1007/978-1-4939-6530-4_9
4. Da Ros, F., Soprano, M., Di Gaspero, L., Roitero, K.: Large language models for combinatorial optimization: a systematic review. arXiv preprint arXiv:2507.03637 (2025)
5. Zou, H.P., et al.: LLM-Based Human-Agent Collaboration and Interaction Systems: A Survey (2025). https://arxiv.org/abs/2505.00753
6. Xia, B., Lu, Q., Zhu, L., Xing, Z., Zhao, D., Zhang, H.: Evaluation-Driven Development and Operations of LLM Agents: A Process Model and Reference Architecture (2025). https://arxiv.org/abs/2411.13768
7. Liu, F., et al.: Large language model for multiobjective evolutionary optimization. In: International Conference on Evolutionary Multi-Criterion Optimization, pp. 178–191 (2025). Springer
8. Sartori, C.C., Blum, C., Bistaffa, F., Corominas, G.R.: Metaheuristics and large language models join forces: Towards an integrated optimization approach. IEEE Access (2024)
9. Li, R., Wang, L., Sang, H., Yao, L., Pan, L.: LLM-assisted automatic memetic algorithm for lot-streaming hybrid job shop scheduling with variable sublots. IEEE Transactions on Evolutionary Computation (2025)
10. Hao, H., Zhang, X., Zhou, A.: Large language models as surrogate models in evolutionary algorithms: a preliminary study. Swarm Evol. Comput. **91**, 101741 (2024)
11. Dat, P.V.T., Doan, L., Binh, H.T.T.: Hsevo: elevating automatic heuristic design with diversity-driven harmony search and genetic algorithm using LLMs. In: Proceedings of the AAAI Conference on Artificial Intelligence, vol. 39, pp. 26931–26938 (2025)
12. Ye, H., et al.: Reevo: large language models as hyper-heuristics with reflective evolution. Adv. Neural. Inf. Process. Syst. **37**, 43571–43608 (2024)
13. Bömer, T., Koltermann, N., Disselnmeyer, M., Dörr, L., Meyer, A.: Leveraging large language models to develop heuristics for emerging optimization problems. arXiv preprint arXiv:2503.03350 (2025)
14. Stein, N.v., Bäck, T.: Llamea: A large language model evolutionary algorithm for automatically generating metaheuristics. Trans. Evol. Comp 29(2), 331–345 (2025) https://doi.org/10.1109/TEVC.2024.3497793
15. Hansen, P., Mladenović, N., Moreno Perez, J.A.: Variable neighbourhood search: methods and applications. Ann. Oper. Res. **175**(1), 367–407 (2010)
16. Costa, L.R., Aloise, D., Mladenović, N.: Less is more: basic variable neighborhood search heuristic for balanced minimum sum-of-squares clustering. Inf. Sci. **415**, 247–253 (2017)
17. Pyatkin, A., Aloise, D., Mladenović, N.: Np-hardness of balanced minimum sum-of-squares clustering. Pattern Recogn. Lett. **97**, 44–45 (2017). https://doi.org/10.1016/j.patrec.2017.05.033
18. Martín-Santamaría, R., Sánchez-Oro, J., Pérez-Peló, S., Duarte, A.: Strategic oscillation for the balanced minimum sum-of-squares clustering problem. Inform. Sci. **585**, 529–542 (2022). https://doi.org/10.1016/j.ins.2021.11.048
19. Mladenović, N., Drezner, Z., Brimberg, J., Urošević, D.: Less is more approach in heuristic optimization. In: The Palgrave Handbook of Operations Research, pp. 469–499. Springer, Cham, Switzerland (2022)

20. Serebryany, K., Bruening, D., Potapenko, A., Vyukov, D.: AddressSanitizer: a fast address sanity checker. In: 2012 USENIX Annual Technical Conference (USENIX ATC 12), pp. 309–318. USENIX Association, Boston, MA (2012)
21. Kelly, M., Longjohn, R., Nottingham, K.: The UCI Machine Learning Repository. https://archive.ics.uci.edu
22. Novikov, A., et al.: AlphaEvolve: a coding agent for scientific and algorithmic discovery (2025). https://arxiv.org/abs/2506.13131

A Variable Neighborhood Search Heuristic for Semi-supervised Minimum Sum-of-Squares Clustering

Mahuton Hugues Midingoyi, André Meneses, and Daniel Aloise[(✉)]

GERAD and Polytechnique Montréal, Montreal H3T 1J4, Canada
`daniel.aloise@polymtl.ca`

Abstract. Semi-supervised clustering is a learning approach that primarily relies on unlabeled data but incorporates some prior information to improve the clustering results. Among various clustering objectives, the minimum sum-of-squares clustering (MSSC) is widely used to partition data by minimizing intra-cluster variances. In our work, we propose a Variable Neighborhood Search (VNS) heuristic for semi-supervised MSSC, where prior information is given in the form of pairwise must-link and cannot-link constraints. Our approach reformulates the optimization problem by representing must-link constraints through the construction of super-points, which implicitly satisfy these constraints, while cannot-link constraints are incorporated as penalties in the objective function. Computational experiments indicate that, in the majority of tested cases, our proposed VNS heuristic outperforms the solutions obtained by the state-of-the-art heuristic algorithm found in the literature within the same computational time.

Keywords: Semi-supervised clustering · pairwise constraints · variable neighborhood search

1 Introduction

The Minimum Sum-of-Squares Clustering (MSSC) problem is central in data analysis. It consists of partitioning a set of data points into a fixed number of clusters by minimizing intra-cluster variances. The problem is NP-hard in general dimension for $k \geq 2$ [3], being the k-means algorithm [12] the most popular heuristic to approach the problem.

Clustering is conventionally an unsupervised learning process. Without guidance from supervision, clustering algorithms may produce results that do not reflect the underlying structure of the data. Consequently, researchers have explored incorporating prior knowledge into the learning process to improve clustering quality. This approach is commonly referred to as semi-supervised clustering or constrained clustering.

S. Cavero et al. (Eds.): ICVNS 2025, LNCS 16256, pp. 166–179, 2026.
https://doi.org/10.1007/978-3-032-19582-1_12

Clustering problems can incorporate various forms of prior knowledge (see Cai et al. [7] for a detailed review). A common example is pairwise constraints, which include must-link and cannot-link constraints. Must-link constraints specify pairs of points that must belong to the same cluster, while cannot-link constraints specify pairs that must belong to different clusters.

Formally, given a set $P = \{p_1, p_2, \ldots p_n\}$ of n data points in $\mathbb{R}^d$, let

$$ML = \{(i,j) \mid p_i \text{ and } p_j \text{ must be assigned to the same cluster}\},$$

$$CL = \{(i,j) \mid p_i \text{ and } p_j \text{ must be assigned to different clusters}\}.$$

denote the sets of must-link and cannot-link pairs, respectively. Then, the semi-supervised MSSC by pairwise constraints (MSSC) can be formulated as the following mixed-integer nonlinear programming (MINLP) model:

$$\min_{x} \quad \sum_{\ell=1}^{k} \frac{\sum_{i=1}^{n-1} \sum_{j=i+1}^{n} x_{i\ell} x_{j\ell} \|p_i - p_j\|_2^2}{\sum_{i=1}^{n} x_{i\ell}} \tag{1a}$$

$$\text{s.t.} \quad \sum_{\ell=1}^{k} x_{i\ell} = 1 \quad \forall i = 1, \ldots, n \qquad , \tag{1b}$$

$$x_{i\ell} = x_{j\ell} \quad \forall \ell = 1, \ldots, k, \ \forall (i,j) \in ML \quad , \tag{1c}$$

$$x_{i\ell} + x_{j\ell} \leq 1 \quad \forall \ell = 1, \ldots, k, \ \forall (i,j) \in CL, \tag{1d}$$

$$x_{i\ell} \in \{0, 1\} \quad \forall i = 1, \ldots, n, \forall \ell = 1, \ldots, k, \tag{1e}$$

where the binary variables $x_{i\ell}$ indicate the assignment of point p_i to cluster ℓ, with $x_{i\ell} = 1$ meaning that p_i belongs to cluster ℓ, and $x_{i\ell} = 0$ otherwise. Constraints (1b) guarantee that each point is assigned to exactly one of the k clusters, whereas constraints (1c) and (1d) enforce the must-link and cannot-link relationships for all pairs in ML and CL, respectively.

In this paper, we propose an efficient Variable Neighborhood Search (VNS) [19] heuristic for the semi-supervised MSSC problem by pairwise constraints. Our approach is based on reformulating (1) so that must-link constraints are satisfied implicitly through the construction of super-points, while cannot-link constraints are incorporated as penalty terms in the objective function. Our main contributions can be summarized as follows:

- We propose a new formulation of the semi-supervised MSSC problem in which must-link constraints are implicitly satisfied through the creation of super-points. This transformation significantly reduces the dimensionality of the problem. The cannot-link constraints are incorporated as penalty terms in the objective function, leading to a penalized optimization model that maintains flexibility while encouraging constraint satisfaction.
- We develop a specialized Variable Neighborhood Search (VNS) heuristic tailored to the penalized semi-supervised MSSC formulation. The method systematically explores multiple neighborhood structures and applies an efficient local search procedure based on incremental cost updates, allowing for rapid move evaluations.

– We conduct extensive experiments on benchmark data instances used in the literature. Our results show that the proposed VNS consistently achieves optimal or near-optimal solutions, outperforming or matching the state-of-the-art heuristic within the same computing times.

The remainder of the paper is organized as follows. Section 2 presents a review of related works for solving the semi-supervised MSSC problem. Section 3 presents our methodology composed by two principal components: a reformulation of the problem described in Sect. 3.1, and our developed VNS method detailed in Sect. 3.2. Computational experiments are reported in Sect. 4, where our method is compared with the current state-of-the-art heuristic found in the literature. Finally, conclusions are given in Sect. 5.

2   Related Works

Semi-supervised MSSC by pairwise constraints has been studied by many authors in the literature. Based on different non-linear mixed-integer programming models, exact methods are developed to guarantee the optimality of the obtained solutions. Among them, we can enumerate the constraint programming approach proposed by Dao et al. [10], the column generation approach developed by Babaki et al. [4] and the branch-and-bound approach of Guns et al. [15]. A Branch-and-Cut algorithm proposed by Piccialli et al. [20] for the semi-supervised MSSC by pairwise constraints is considered to be the state-of-the-art method. This algorithm enables the resolution of problem instances with up to 800 data points in high dimensions, and for 8 clusters within a reasonable running time.

Due to the large number of data points in practice, heuristic algorithms have been designed for the problem, such that local or near optimal solutions are obtained. To the best of our knowledge, the first work to propose a heuristic (Cop-KMeans) for the problem was presented in [24] where the authors modified the classical k-means algorithm for (unsupervised) MSSC in order to make use of prior information. Later, other heuristic methods proposed enhancements to the Cop-Kmeans heuristic (e.g. [17,22,23,26]).

Using a Lagrangian framework, Ganji et al. [13] address the semi-supervised MSSC problem by aggregating points connected by must-link constraints into super-points, thereby ensuring that all must-link constraints are satisfied. Their algorithm is based on an assignment approach and, in each iteration, includes a procedure that moves points between clusters with the goal of minimizing violations of cannot-link constraints. Regarding metaheuristics for the semi-supervised MSSC by pairwise constraints, González-Almagro et al. [14] proposed a dual Iterative Local Search (ILS), which introduces diversity to basic ILS framework.

More recently, Baumann et al. [5] proposed a heuristic called PCCC for semi-supervised MSSC, which allows pairwise constraints to be treated as either *hard*, i.e., requiring strict satisfaction, or *soft*, i.e., permitting violations that incur a penalty. The algorithm iteratively assigns data points to cluster centers while respecting the pairwise constraints and updates the cluster centers at each iteration. The assignment step of data points to centers is performed by solving a binary program. PCCC is capable of producing optimal or near-optimal solutions for large benchmark instances and is considered state-of-the-art for the problem.

In this work, we choose to implement a VNS heuristic for the problem, given its previous success in various other clustering problems found in the literature [6,8,9,16,25]. In particular, our method reformulates the semi-supervised MSSC with pairwise constraints (1) as an unconstrained problem, where must-link constraints are enforced through variable elimination, and cannot-link constraints are incorporated via large penalty coefficients in the objective function. The details of our approach are presented in the next section.

3   Methodology

The proposed methodology consists of two principal components. The first involves the reformulation of model (1) as an equivalent MINLP without explicit pairwise constraints, whereas the second implements a variable neighborhood search heuristic to address the problem.

3.1   Problem Reformulation

Our problem reformulation is structured around two main steps: problem size reduction and penalization, which respectively address the treatment of must-link and cannot-link constraints.

Problem Size Reduction. The must-link constraints enable a reduction in the number of variables in (1), thereby eliminating constraints (1c) from the formulation.

More in detail, we construct an undirected graph from the obtained set of must-link constraints where each node is a data point, and there is an edge between two nodes if the corresponding data points are involved in a must-link constraint, i.e., belong to the same cluster. Next, we compute the transitive closure of the graph leading to a partition $B_1, \ldots, B_{\bar{n}}$ of P into $\bar{n} < n$ components for ML non-empty. We represent these components as super-points $\{\bar{p}_i\}_{i=1}^{\bar{n}}$, where each super-points $\bar{p}_i$ is associated to a weight w_i equal to the number of points in the component. These super-points can be used to reconstruct the original data points in P, and conversely, if the set ML is known.

To perform data clustering with these super-points, we define a dissimilarity measure D_{ij} between the super-points $\bar{p}_i$ and $\bar{p}_j$ as follows:

$$D_{ij}^2 = \begin{cases} \displaystyle\sum_{(s,t)\in B_i \times B_j} \|p_s - p_t\|^2, & \text{if } i \neq j, \\ \displaystyle\sum_{\substack{s,t\in B_i \\ s<t}} \|p_s - p_t\|^2, & \text{if } i = j. \end{cases} \tag{2}$$

Therefore, problem (1) can be reformulated as:

$$\min_{\bar{x}} \quad \sum_{\ell=1}^{k} \frac{\sum_{i=1}^{\bar{n}} \sum_{j=i}^{\bar{n}} D_{ij}^2 \bar{x}_{i\ell} \bar{x}_{j\ell}}{\sum_{i=1}^{\bar{n}} w_i \bar{x}_{i\ell}} \tag{3a}$$

$$\text{s.t.} \quad \sum_{\ell=1}^{k} \bar{x}_{i\ell} = 1 \quad \forall i = 1, \ldots, \bar{n}, \tag{3b}$$

$$\bar{x}_{i\ell} + \bar{x}_{j\ell} \le 1 \quad \forall \ell = 1, \ldots, k, \ \forall (i,j) \in \overline{CL}, \tag{3c}$$

$$\bar{x}_{i\ell} \in \{0,1\} \quad \forall i = 1, \ldots, \bar{n}, \ \forall \ell = 1, \ldots, k, \tag{3d}$$

where, $\bar{x}_{i\ell} = 1$ indicates that super-point $\bar{p}_i$ is assigned to cluster ℓ, $\bar{x}_{i\ell} = 0$ otherwise. Note that constraints (1c) no longer appears explicitly in model (3), as it is now implicitly satisfied.

The idea of using must-link constraints to reduce the dimensionality of clustering problems is not new in the literature and has been used by diverse methods that optimize the MSSC criterion under pairwise constraints (see, e.g., [11,13,15,20,21]).

Penalization. The set of cannot-link constraints must be adjusted for the super-points. We denote by $\overline{CL}$ the set of index pairs of super-points linked by cannot-link constraints. The set $\overline{CL}$ consists of all pairs of super-point indices (i,j) such that there exists (s,t) in CL with $s \in B_i$ and $t \in B_j$.

We opted to incorporate the cannot-link constraints as a penalty in the objective function, allowing us to account for them without explicitly including them in the problem formulation. This approach gives our heuristic method greater flexibility by eliminating the need to check constraint satisfaction during the search. Thus, each cannot-link constraint violation is penalized in the objective function using the following penalized metric $\bar{D}$, where:

$$\bar{D}_{ij} = \begin{cases} \infty & \text{if } (i,j) \in \overline{CL}, \\ D_{ij} & \text{otherwise.} \end{cases} \tag{4}$$

In practical implementation, the infinity value is replaced by $M = (n-1) \times d_{max}$, where d_{max} corresponds to the maximum distance between a pair of points in the dataset. This choice of M ensures that it exceeds the cost of any feasible solution of the problem.

Finally, the following reformulation is obtained for the semi-supervised MSSC by pairwise constraints:

$$\min_{\bar{x}} \quad \sum_{\ell=1}^{k} \frac{\sum_{i=1}^{\bar{n}} \sum_{j=i}^{\bar{n}} \bar{D}_{ij}^2 \bar{x}_{i\ell} \bar{x}_{j\ell}}{\sum_{i=1}^{\bar{n}} w_i \bar{x}_{i\ell}} \tag{5a}$$

$$\text{s.t.} \quad \sum_{\ell=1}^{k} \bar{x}_{i\ell} = 1 \quad \forall i = 1, \ldots, \bar{n} \quad , \tag{5b}$$

$$\bar{x}_{i\ell} \in \{0, 1\} \quad \forall i = 1, \ldots, \bar{n}, \ \forall \ell = 1, \ldots, k \tag{5c}$$

We can notice that pairwise constraints (1c) and (1d) are not explicitly enforced in (5), which is the underlying optimization problem approached by our VNS heuristic explained in the next section.

3.2 VNS Algorithm for Pairwise Constrained MSSC

In this section, we present the key aspects of our VNS heuristic for semi-supervised MSSC by pairwise constraints. Our VNS leverages multiple neighborhood structures and a local search to explore the solution space efficiently.

VNS is a metaheuristic for combinatorial optimization that systematically changes neighborhoods to escape local optima and enhance global exploration [19]. It has been successfully applied to several clustering problems [1,2,6,9,16,21].

Starting from an initial solution x, the algorithm iteratively performs the following steps. First, a solution x' is generated by perturbing x within the current neighborhood (*shaking*). A local search is then applied to x', yielding a locally optimal solution x''. If x'' improves upon x, the incumbent solution is updated ($x \leftarrow x''$) and the search restarts from the first neighborhood; otherwise, the algorithm proceeds to the next neighborhood in the sequence. The process continues until a stopping criterion, such as a time limit or absence of improvement, is met.

Neighborhood Evaluation. Our VNS heuristic works in its shaking by relocating t points from their current cluster to a different one. Accordingly, the neighborhood structure $\mathcal{N}_t$ of a solution consists of all configurations obtained by moving t points to alternative clusters.

Consider super-points partitioned into k clusters $C_1, C_2, \ldots, C_k$, forming a solution $\mathcal{C}$. The contribution of cluster C_ℓ to the objective function of (5) can be expressed as:

$$\text{Cost}(C_\ell) = \frac{\sum_{i=1}^{\bar{n}} \sum_{j=i}^{\bar{n}} \bar{D}_{ij}^2 \bar{x}_{i\ell} \bar{x}_{j\ell}}{\sum_{i=1}^{\bar{n}} w_i \bar{x}_{i\ell}} \tag{6}$$

or equivalently by:

$$\text{Cost}(C_\ell) = \frac{\bar{D}_\ell}{s_\ell} \tag{7}$$

where $s_\ell = \sum_{i:\bar{p}_i \in C_\ell} w_i$ denotes the number of data points in cluster C_ℓ, and $\bar{\mathcal{D}}_\ell = \sum_{i<j:\bar{p}_i,\bar{p}_j \in C_\ell} \bar{D}_{ij}^2$ corresponds to the sum of the dissimilarities between the super-points in cluster C_ℓ.

Note that the numerator of the cluster cost is a quadratic term. This observation allows for an efficient computation of the cost variation in (6) when adding or removing a super-point to or from a cluster in the partition. The variation γ_ℓ^i in the numerator of the cost of C_ℓ resulting from moving the super-point $\bar{p}_i$ to it is given by:

$$\gamma_\ell^i = \sum_{j:\bar{p}_j \in C_\ell} \bar{D}_{ij}^2, \tag{8}$$

while the change in the denominator corresponds to the weight of the point being added or removed. Thus, the cost of adding super-point $\bar{p}_i$ to cluster C_ℓ is computed as:

$$\frac{\bar{\mathcal{D}}_\ell + \gamma_\ell^i}{s_\ell + w_i}, \tag{9}$$

while the cost of removing a super-point $\bar{p}_i$ from cluster C_r is computed as:

$$\frac{\bar{\mathcal{D}}_r - \gamma_r^i}{s_r - w_i}. \tag{10}$$

Once a super-point $\bar{p}_i$ is moved in solution $\mathcal{C}$ from cluster C_r to another cluster C_ℓ, all variation values $\gamma_\ell^j, \gamma_r^j$ for every $j = 1, \ldots, \bar{n}, j \neq i$ are directly updated as:

$$\gamma_r^j \leftarrow \gamma_r^j - \bar{D}_{ij}^2 \text{ and } \gamma_\ell^j \leftarrow \gamma_\ell^j + \bar{D}_{ij}^2. \tag{11a}$$

Consequently, move evaluations in our neighborhood are performed in constant time, while updates of auxiliary structures γ are performed in $O(\bar{n})$.

Local Search. Our local search is based on analyzing the possibilities of moving one point from one cluster to another, i.e., $\mathcal{N}_1$. Since identifying the best improving move can be computationally expensive, our local search adopts a compromise between the best- and first-improvement strategies. It evaluates each pair of clusters to find the best improving move. Let us denote:

– $\mathcal{K}$: the set of all combinations of two distinct cluster indices, defined as:

$$\mathcal{K} = \{(i,j) \mid 1 \leq i < j \leq k\}; \tag{12}$$

– $\mathcal{K}_l$: the subset of all pairs in $\mathcal{K}$ that include index l, for $l = 1, \ldots, k$. It is given by:

$$\mathcal{K}_l = \{(i,j) \mid (i,l) \in \mathcal{K} \vee (l,j) \in \mathcal{K}\}. \tag{13}$$

The pseudo-code presented in Algorithm 1 describes our local search procedure.

Algorithm 1. Local Search

1: **Input:** Initial solution $\mathcal{C}^0$
2: **Output:** Local solution $\mathcal{C}$
3: Initialize $\mathcal{C} \leftarrow \mathcal{C}^0$
4: Initialize $\texttt{next\_}\mathcal{K} \leftarrow \mathcal{K}$
5: **while** True **do**
6:      $\texttt{current\_}\mathcal{K} \leftarrow \texttt{next\_}\mathcal{K}$
7:      $\texttt{next\_}\mathcal{K} \leftarrow \emptyset$
8:      **for** each (i, j) in $\texttt{current\_}\mathcal{K}$ **do**
9:           Find the best move of a super-point between the clusters C_i and C_j
10:           **if** the found move between C_i and C_j improves the objective value **then**
11:                Apply the move and update $\mathcal{C}$
12:                $\texttt{next\_}\mathcal{K} \leftarrow \texttt{next\_}\mathcal{K} \cup \mathcal{K}_i \cup \mathcal{K}_j$
13:           **end if**
14:      **end for**
15:      **if** $\texttt{next\_}\mathcal{K}$ is empty **then**
16:           **return** the solution $\mathcal{C}$
17:      **end if**
18: **end while**

Algorithm 1 starts with a current solution $\mathcal{C}$, initialized from an initial solution $\mathcal{C}^0$ (line 3). The algorithm initializes a set $\texttt{next\_}\mathcal{K}$ at line 4 containing all possible pairs of cluster indices for exploration. Then, the algorithm starts an iterative loop (lines 5–18) that continues as long as improving moves of super-points can be found. At each iteration, the set $\texttt{current\_}\mathcal{K}$ receives the cluster index pairs to be tested from the previous iteration (line 6). An initially empty set $\texttt{next\_}\mathcal{K}$ is prepared then in line 7 to store the pairs to be reconsidered in the next iteration. In the sequel, each pair (i, j) of clusters in $\texttt{current\_}\mathcal{K}$ is examined in lines 8-14. The best potential move of a super-point in clusters C_i or C_j is found in 9. If several moves have the best objective value, only one move is considered among them. If it is detected that the considered super-point move between clusters C_i and C_j improve the objective value in line 10, the move is performed in line 11. Updating the current solution involves updating the size and cost of each affected cluster, as well as the variation values, using (8)–(11). Additionally, all cluster index pairs involving one of the two modified clusters are added to $\texttt{next\_}\mathcal{K}$ in line 12 to be reconsidered in the next iteration. This mechanism avoids a full exploration of all cluster pairs at each iteration, focusing the search only on solution components directly affected by the solution modifications. The main loop continues until no cluster pairs are left to explore in the next iteration (lines 15–17). In that case, a local minimum is returned in line 16.

3.3   Feasibility Step

The VNS process does not guarantee that the final solution satisfies all cannot-link constraints. Therefore, a post-processing step is applied whenever the method produces an infeasible solution for the semi-supervised MSSC.

Let $\mu_1, \ldots, \mu_k$ denote the centroids of the final solution $\mathcal{C} = \{C_1, \ldots, C_k\}$ returned by the VNS heuristic. The feasibility step consists in solving the following MIP:

$$\min_{\bar{x}} \quad \sum_{\ell=1}^{k} \sum_{i=1}^{\bar{n}} \bar{x}_{i\ell} \|\bar{z}_i - \mu_\ell\|^2 \tag{14a}$$

$$\text{s.t.} \quad \sum_{\ell=1}^{k} \bar{x}_{i\ell} = 1 \quad \forall i = 1, \ldots, n \tag{14b}$$

$$\bar{x}_{i\ell} + \bar{x}_{j\ell} \leq 1 \quad \forall \ell = 1, \ldots, k, \; \forall (i,j) \in \overline{CL}, \tag{14c}$$

$$\bar{x}_{i\ell} \in \{0,1\} \quad \forall i = 1, \ldots, n, \; \forall \ell = 1, \ldots, k. \tag{14d}$$

where $\bar{z}_i$ is the centroid of the points merged at super-point $\bar{p}_i$, for $i = 1, \ldots, \bar{n}$.

4   Computational Experiments

To evaluate the performance of our VNS heuristic[1], we used the COL1 benchmark suite of 25 datasets from [5]. These datasets contain between 47 and 846 samples, span 2 to 90 dimensions, and include 2 to 15 clusters. The corresponding sets of pairwise constraints were also obtained from [5], with each set defining a distinct instance of the semi-supervised MSSC problem. For each dataset, the authors constructed four different configurations of pairwise constraints, resulting in a total of $25 \times 4 = 100$ data instances.

The computational experiments were performed on a Intel 2.5Ghz processor with 24GB of RAM memory and compiled by gcc 11.4. Both algorithms were executed 10 times for the same amount of CPU time (determined by the average convergence time of the PCCC implementation available in [5]). We note that, although PCCC is implemented in Python, it primarily relies on Gurobi (version 12.0.3) to iteratively solve its auxiliary assignment steps.

The ten executions of the VNS and PCCC algorithms are paired, as each begins from the same initial solution generated by the classical k-means algorithm [18]. This initialization is performed on the super-points described in Sect. 3.1, where each super-point is represented by the average coordinates of its component data points, thereby respecting the must-link constraints. The VNS parameters were set to $t_{max} = \min\{200, 0.75 \times \bar{n}\}$ with step increments of $\min\{20, 0.1 \times t_{max}\}$.

We present the results for the COL1 data instances in two separate tables. Table 1 reports the cases for which optimal solutions are known using the algorithm of [20]. For these instances, we provide the average solution gaps relative to the corresponding optimal values. The table also includes, for each data instance, the number of original data points (n), the number of clusters (k),

---

[1] available in C++ at https://github.com/andre-meneses/Constraint_Clustering_VNS.

Table 1. Comparison between our VNS and PCCC in COL1 instances with known optimal values.

Dataset	n	k	ML	CL	VNS		PCCC		time
					cost	gap (%)	cost	gap (%)	
Appendicitis	106	2	13	2	491.80	**0.00**	493.91	0.43	0.09
			39	16	544.18	**0.55**	545.14	0.73	0.04
			71	49	612.93	**0.00**	612.93	**0.00**	0.04
			164	67	612.93	**0.00**	612.93	**0.00**	0.01
Breast Cancer	5300	2	216	190	12139.45	0.46	12084.17	**0.00**	0.20
			876	720	12185.03	**0.00**	12185.03	**0.00**	0.02
Bupa	345	2	79	74	1853.13	**0.17**	1853.79	0.21	0.21
			323	272	2041.64	**0.00**	2041.93	0.01	0.04
Circles	300	2	50	55	473.20	**0.01**	480.76	1.61	0.23
			208	227	598.80	**0.00**	599.08	0.05	0.05
Ecoli	336	8	357	918	1027.69	**0.00**	1034.12	0.63	1.00
			609	1669	1111.78	**0.00**	1120.20	0.76	0.26
Glass	214	6	11	44	812.41	**0.00**	842.19	3.67	0.57
			139	389	1266.22	**0.00**	1314.49	3.81	0.66
			259	644	1409.89	**0.00**	1416.29	0.45	0.36
Haberman	306	2	76	44	799.45	**0.03**	802.68	0.43	0.24
			304	161	889.43	**0.00**	894.97	0.62	0.05
			634	401	891.42	**0.00**	891.42	**0.00**	0.02
Hayes-roth	160	3	12	16	446.67	**0.00**	453.33	1.49	0.22
			102	174	553.10	**0.00**	553.48	0.07	0.13
			177	319	551.07	**0.00**	552.83	0.32	0.08
Heart	270	2	41	50	3044.98	**0.00**	3118.20	2.40	0.30
			178	173	3085.12	**0.00**	3287.93	6.57	0.06
Ionosphere	351	2	92	61	10265.88	0.07	10260.34	**0.01**	0.24
			330	300	10944.90	**0.00**	10946.22	0.01	0.06
Iris	150	3	6	22	146.30	**0.00**	146.30	**0.00**	0.15
			25	80	146.15	**0.00**	146.28	0.09	0.18
			57	196	140.97	**0.00**	141.22	0.18	0.25
			94	341	145.01	**0.00**	145.11	0.07	0.14
Led7digit	500	10	267	2508	1449.42	**0.00**	1449.99	0.04	6.43
			460	4490	1511.03	**0.00**	1511.19	0.01	2.12
Monk-2	432	2	101	130	2389.91	1.32	2358.87	**0.00**	0.41
			473	473	2382.69	**0.00**	2382.69	**0.00**	0.06
Moons	300	2	55	50	283.74	**0.00**	283.75	**0.00**	0.16
			200	235	322.11	**0.00**	322.27	0.05	0.07
			494	496	322.93	**0.00**	322.93	**0.00**	0.02
Mov. Libras	360	15	6	147	10597.90	**0.00**	11016.47	3.95	2.21
Newthyroid	215	3	25	30	506.06	**0.00**	506.06	**0.00**	0.24
			108	123	536.00	**0.00**	536.31	0.06	0.18
			270	258	550.93	**0.00**	550.93	**0.00**	0.08
			449	454	550.36	**0.00**	550.76	0.07	0.05
Saheart	462	2	152	124	3795.75	1.04	3757.36	**0.02**	0.28
			595	486	3924.31	**0.00**	3926.08	0.05	0.04
Sonar	208	2	29	26	11289.27	**0.01**	11308.61	0.18	0.14
			100	110	11873.80	**0.00**	11873.92	**0.00**	0.10
			245	251	11962.93	**0.00**	11962.93	**0.00**	0.03
Soybean	47	4	0	3	367.14	**0.00**	372.77	1.53	0.05
			4	6	367.14	**0.00**	369.16	0.55	0.05
			6	22	367.14	**0.00**	367.14	**0.00**	0.07
			12	33	367.14	**0.00**	367.14	**0.00**	0.06
Spectfheart	267	2	56	35	10467.75	**0.00**	10472.92	0.05	0.22
			233	118	11210.50	**0.00**	11488.33	2.48	0.08
			543	277	11268.30	**0.00**	11268.30	**0.00**	0.03
Spiral	300	2	52	53	462.24	**0.15**	462.51	0.21	0.20
			224	211	558.25	**0.00**	558.92	0.12	0.05
Tae	151	3	8	20	478.85	**0.00**	490.89	2.51	0.22
			82	171	684.30	**0.00**	685.93	0.24	0.15
			151	314	711.83	**0.00**	711.83	**0.00**	0.07
Vehicle	846	4	874	2696	13303.38	0.40	13291.72	**0.31**	0.33
			1955	6046	13334.20	**0.00**	13334.20	**0.00**	0.06
Wine	178	3	8	28	1284.69	**0.00**	1284.87	0.01	0.13
			31	122	1285.13	**0.00**	1285.13	**0.00**	0.21
			70	281	1285.13	**0.00**	1285.13	**0.00**	0.22
			143	487	1290.66	**0.00**	1291.36	0.05	0.11
Zoo	101	7	7	8	538.71	**0.00**	555.24	3.07	0.18
			21	34	545.96	**0.00**	565.45	3.57	0.23
			29	91	551.82	**0.00**	567.44	2.83	0.27
			41	169	552.75	**0.00**	560.39	1.38	0.33

Table 2. Comparison between VNS and PCCC in COL1 instances when optimal solution values are not known.

Dataset	n	k	ML	CL	VNS	PCCC	time
Breast Cancer	5300	2	1965	1690	**12214.60**	**12214.60**	0.01
Bupa	345	2	699	627	**2047.30**	**2047.30**	0.01
			1201	1145	**2047.30**	**2047.30**	0.01
Circles	300		502	488	**599.99**	**599.99**	0.01
Ecoli	336	8	30	106	**791.15**	826.26	1.97
			163	398	**944.88**	960.64	1.42
Glass	214	6	52	179	**1085.60**	1110.41	0.80
Haberman	306	2	1135	756	**891.42**	**891.42**	0.01
Hayes-roth	160	3	39	81	**513.81**	516.21	0.31
Heart	270	2	396	424	**3120.52**	3302.53	0.02
Ionosphere	351	2	732	646	**10971.50**	**10971.50**	0.02
Led7digit	500	10	25	275	**1139.61**	1183.66	1.54
			126	1099	**1352.09**	1381.96	4.58
Monk-2	432	2	979	1101	**2384.29**	**2384.29**	0.01
Moons	300	2	900	870	**322.93**	**322.93**	0.01
Movement Libras	360	15	27	603	**11508.30**	11746.60	5.95
			112	1319	**15216.75**	15548.58	10.48
			158	2398	**17956.65**	18306.36	19.39
Saheart	462	2	1292	1123	**3927.71**	**3927.71**	0.02
Sonar	208	2	436	425	**11962.93**	**11962.93**	0.01
Spectfheart	267	2	965	466	**11268.30**	**11268.30**	0.01
Spiral	300	2	487	503	**564.47**	**564.47**	0.01
			918	852	**564.47**	**564.47**	0.01
Tae	151	3	40	80	**594.48**	596.27	0.33
Vehicle	846	4	221	682	9957.60	**9879.45**	2.36

the number of must-link (ML) and cannot-link (CL) constraints, along with the average CPU time (in seconds) allocated for each algorithm run.

We observe from Table 1 that our VNS obtained the optimal solutions in 57 out of 68 instances (approximately 84%). In the remaining 11 instances, the observed gaps were very small, ranging from 0.01% to 1.32%, with an average of 0.38%, thereby confirming the high effectiveness and stability of the heuristic. For comparison, PCCC reached the optimal value in 21 out of 68 instances (approximately 31%), while the remaining cases showed small deviations with an average gap of approximately 1.02%.

Table 2 reports average solution values for the remaining COL1 data instances. We note that, for some instances, the number of super-points $\bar{n}$ was equal to the number of clusters k after processing the must-link constraints. Since these instances are trivially solved by assigning each super-point to a singleton cluster, we excluded them from our comparison with the PCCC algorithm.

We note from Table 2 that, overall, our VNS consistently produced equal or lower average costs than PCCC, except for the Vehicle instance. Across all 25 instances, VNS achieved an average relative improvement of approximately 1.02% over PCCC. The largest relative improvement, 5.51%, occurred on the Heart dataset instance.

Finally, it is worth noting that our VNS consistently produced final solutions that satisfied all cannot-link constraints, except for one instance of the Vehicle dataset (ML = 874, CL = 2696). For this instance, the feasibility step described in Sect. 3.3 was performed. The time spent by Gurobi on solving the MIP (14) was 0.06 s in average.

5   Conclusions

In this work, we proposed a Variable Neighborhood Search (VNS) heuristic for the semi-supervised Minimum Sum-of-Squares Clustering (MSSC) problem with pairwise must-link and cannot-link constraints. The proposed method addresses the limitations of existing approaches by reformulating the problem to reduce its size and simplify constraint handling. Specifically, must-link relationships are implicitly satisfied through the construction of super-points, thereby reducing the dimensionality of the problem, while cannot-link relationships are incorporated into the objective function as penalty terms. This leads to a penalized optimization formulation in which cannot-link relationships are handled implicitly within the objective function, providing a more tractable structure for metaheuristic search.

The VNS heuristic was designed with multiple complementary neighborhood structures and an efficient local search mechanism that exploits incremental cost updates for constant-time move evaluation. Additionally, a feasibility restoration step based on mixed-integer programming was integrated to guarantee full satisfaction of cannot-link constraints in the final solution when necessary. Together, these design elements enable the algorithm to efficiently explore the search space while maintaining solution feasibility. Computational experiments on benchmark data instances demonstrated that the proposed VNS heuristic consistently delivers high-quality clustering results. Across 68 benchmark instances with known optimal solutions, the algorithm attained the optimum in approximately 84% of cases and showed very small average deviations (below 0.4%) on the remaining instances. Overall, when compared to the state-of-the-art PCCC algorithm, our method achieved comparable or superior performance.

Acknowledgments. The authors gratefully acknowledge the financial support of the Natural Sciences and Engineering Research Council of Canada (NSERC) (grant number 2023-04466).

Disclosure of Interests. The authors have no competing interests to declare that are relevant to the content of this article.

References

1. Alguwaizani, A., Hansen, P., Mladenović, N., Ngai, E.: Variable neighborhood search for harmonic means clustering. Appl. Math. Model. **35**(6), 2688–2694 (2011)
2. Aloise, D., Caporossi, G., Hansen, P., Liberti, L., Perron, S., Ruiz, M.: Modularity maximization in networks by variable neighborhood search. Graph Partitioning Graph Clustering **588**(113) (2012)
3. Aloise, D., Deshpande, A., Hansen, P., Popat, P.: Np-hardness of euclidean sum-of-squares clustering. Mach. Learn. **75**(2), 245–248 (2009)
4. Babaki, B., Guns, T., Nijssen, S.: Constrained clustering using column generation. In: Simonis, H. (ed.) CPAIOR 2014. LNCS, vol. 8451, pp. 438–454. Springer, Cham (2014). https://doi.org/10.1007/978-3-319-07046-9_31
5. Baumann, P., Hochbaum, D.: An algorithm for clustering with confidence-based must-link and cannot-link constraints. INFORMS J. Comput. **37**(4), 1044–1068 (2025). https://doi.org/10.1287/ijoc.2023.0419.cd
6. Brimberg, J., Mladenović, N., Todosijević, R., Urošević, D.: Solving the capacitated clustering problem with variable neighborhood search. Ann. Oper. Res. **272**(1), 289–321 (2019)
7. Cai, J., He, L., Hu, X., Jiang, M., Yu, P.S.: A review on semi-supervised clustering. IEEE Trans. Knowl. Data Eng. **33**(6), 2210–2229 (2021)
8. Carrizosa, E., Mladenović, N., Todosijević, R.: Variable neighborhood search for minimum sum-of-squares clustering on networks. Eur. J. Oper. Res. **230**(2), 356–363 (2013)
9. Costa, L.R., Aloise, D., Mladenović, N.: Less is more: basic variable neighborhood search heuristic for balanced minimum sum-of-squares clustering. Inf. Sci. **415**, 247–253 (2017)
10. Dao, T.B.H., Duong, K.C., Vrain, C.: A declarative framework for constrained clustering. In: Blockeel, H., Kersting, K., Nijssen, S., Železný, F. (eds.) Machine Learning and Knowledge Discovery in Databases: European Conference, ECML PKDD 2013. Lecture Notes in Computer Science, vol. 8190, pp. 419–434. Springer, Berlin, Heidelberg (2013)
11. Du Merle, O., Hansen, P., Jaumard, B., Mladenovic, N.: An interior point algorithm for minimum sum-of-squares clustering. SIAM J. Sci. Comput. **21**(4), 1485–1505 (1999)
12. Forgy, E.W.: Cluster analysis of multivariate data: efficiency versus interpretability of classifications. Biometrics **21**, 768–769 (1965)
13. Ganji, M., Bailey, J., Stuckey, P.J.: Lagrangian constrained clustering. In: Proceedings of the 2016 SIAM International Conference on Data Mining (SDM), pp. 288–296. SIAM (2016)
14. González-Almagro, G., Luengo, J., Cano, J.R., García, S.: Dils: constrained clustering through dual iterative local search. Comput. Operat. Res. **121**, 104979 (2020)
15. Guns, T., Dao, T.B.H., Vrain, C., Duong, K.D.C.: Repetitive branch-and-bound using constraint programming for constrained minimum sum-of-squares clustering. Constraints **21**(3), 364–392 (2016)
16. Hansen, P., Ruiz, M., Aloise, D.: A vns heuristic for escaping local extrema entrapment in normalized cut clustering. Pattern Recogn. **45**(12), 4337–4345 (2012)

17. Huang, H., Cheng, Y., Zhao, R.: A semi-supervised clustering algorithm based on must-link set. In: Tang, C., Ling, C.X., Zhou, X., Cercone, N.J., Li, X. (eds.) ADMA 2008. LNCS (LNAI), vol. 5139, pp. 492–499. Springer, Heidelberg (2008). https://doi.org/10.1007/978-3-540-88192-6_48
18. Lloyd, S.P.: Least squares quantization in pcm. IEEE Trans. Inf. Theory **28**(2), 129–137 (1982)
19. Mladenović, N., Hansen, P.: Variable neighborhood search. Comput. Oper. Res. **24**(11), 1097–1100 (1997)
20. Piccialli, V., Russo, A., Sudoso, A.M.: An exact algorithm for semi-supervised minimum sum-of-squares clustering. Comput. Operat. Res. **147**, 105958 (2022)
21. Randel, R., Aloise, D., Mladenović, N., Hansen, P.: On the k-medoids model for semi-supervised clustering. In: International Conference on Variable Neighborhood Search, pp. 13–27. Springer (2018)
22. Rutayisire, T., Yang, Y., Lin, C., Zhang, J.: A modified cop-Kmeans algorithm based on sequenced cannot-link set. In: Yao, J.T., Ramanna, S., Wang, G., Suraj, Z. (eds.) RSKT 2011. LNCS (LNAI), vol. 6954, pp. 217–225. Springer, Heidelberg (2011). https://doi.org/10.1007/978-3-642-24425-4_30
23. Tan, W., Yang, Y., Li, T.: An improved cop-kmeans algorithm for solving constraint violation. In: Computational Intelligence: Foundations and Applications. World Scientific Proceedings Series on Computer Engineering and Information Science, vol. 4, pp. 690–696. World Scientific Publishing (2010)
24. Wagstaff, K., Cardie, C., Rogers, S., Schrödl, S.: Constrained k-means clustering with background knowledge. In: Proc. 18th Int. Conf. Mach. Learn. (ICML), pp. 577–584. Morgan Kaufmann, San Francisco (2001)
25. Xiao, Y., Huang, C., Huang, J., Kaku, I., Xu, Y.: Optimal mathematical programming and variable neighborhood search for k-modes categorical data clustering. Pattern Recogn. **90**, 183–195 (2019)
26. Yang, Y., Tan, W., Li, T., Ruan, D.: Consensus clustering based on constrained self-organizing map and improved cop-kmeans ensemble in intelligent decision support systems. Knowl.-Based Syst. **32**, 101–115 (2012)

Multi-objective Variable Neighborhood Descent for the Inference of Test Models from User Bug Reports in Software Systems

Alejandro Aunión, Javier Yuste[✉], and Eduardo G. Pardo

Departamento de Informática y Estadística, Universidad Rey Juan Carlos, C/ Tulipán s/n, Móstoles 28933, Spain
{alejandro.aunion,javier.yuste,eduardo.pardo}@urjc.es

Abstract. Software testing activities are critical for the development of high-quality software systems. In this context, high-quality models that accurately represent the system under test are an essential tool. To improve their quality, the inference of these models has been addressed as a multi-objective optimization problem in the literature. In this work, we propose a method based on the Multi-Objective Variable Neighborhood Descent (MO-VND) scheme to solve the problem of inferring behavioral models of software systems built on user-reported bugs. To configure the MO-VND we introduce five different neighborhood structures. The performance of the method is evaluated on a benchmark of real-world instances. The results show that the order in which the neighborhoods are explored greatly affects the performance of the MO-VND method. We compare the proposed method with three well-known algorithms that have already been studied in the literature for this problem: NSGA-II, NSGA-III, and MOEA/D. The results show that, although the proposed MO-VND method is capable of finding high-quality models, there is still room for improvement. In particular, the method might benefit from strategies for an efficient evaluation of neighbor solutions or the identification of promising solutions within the proposed neighborhoods.

Keywords: Model-Based Testing · Model inference · Multi-objective optimization · Heuristics · Multi-Objective Variable Neighborhood Search

1 Introduction

The development of software systems is a complex endeavor, and it is usually performed following a well-defined structure. The Software Development Life-Cycle (SDLC) is a framework that defines different phases, their order, and the transition criteria among them [10]. Generally, SDLCs include activities related to requirements definition, analysis and design, software coding, software testing, and software maintenance. Following an SDLC model during the software development process has been proved to boost the quality of software products [14].

Contrary to common belief, maintenance is usually the most costly phase within the life cycle of software systems, accounting for 60% to 80% of the total development cost [3]. Moreover, software maintenance is a critical activity to preserve the ability of a system to provide the services for which it was designed [10]. It includes tasks such as fixing bugs, improving performance, adapting to new environments, and adding new features. Interestingly, understanding the code is the main driver for such an elevated cost, especially for large projects where there are changes in the software development team [13].

In a similar manner, a correct understanding of the code is crucial for software testing activities. In this context, models that represent the behavior of the system are a fundamental tool for code understanding and the design of quality tests. Poorly designed tests can lead to undetected defects, incomplete fixes, or even new bugs in the system [15].

In software engineering, Model-Based Testing (MBT) refers to the set of activities aimed at designing test cases from models of the system, either manually, automatically or semi-automatically [2]. There exist different types of models that can be useful for these purposes. One of those types is behavioral models, which represent sequences of activities or behaviors of the system [6]. These models are usually defined using state machines. Regardless of the specific type of model used, their quality is crucial for MBT activities [18].

Search-Based Software Engineering (SBSE) is a research field which focuses on defining software engineering tasks as optimization problems and designing and implementing algorithms to solve them [9]. In the context of MBT, a problem arises that can be tackled as an optimization problem: the inference of high-quality models. Several authors have proposed the use of metaheuristics to automatically infer behavioral models [18]. Similarly, in a recent work, some authors proposed an approach to automatically generate behavioral models from bug errors reported by users [8]. They defined the latter as a multi-objective optimization problem, where the quality of models is evaluated using three distinct objectives (see Sect. 2). Finally, they proposed the use of three well-known methods from the literature: NSGA-II [5], NSGA-III [4], and MOEA/D [17].

This work focuses on the behavioral model inference problem introduced by Guizzo *et al.* [8]. Five different neighborhood structures are proposed and analyzed (see Sect. 3). Then, they are combined in a method based on the Multi-Objective Variable Neighborhood Descent scheme. The proposed methods are empirically validated and compared with the state of the art (see Sect. 4). The conclusions and future works are discussed in Sect. 5.

2   Problem Definition

In this work, we study the problem of behavioral model inference from user reports, introduced by Guizzo *et al.* [8]. In this problem, an instance is defined by a collection of user-reported sequences of activities. Each sequence of activities is known as a trace (w). Therefore, an instance is formed by a set of traces $W = \{w_1, w_2, \ldots, w_n\}$.

The goal of the problem is to find a behavioral model (m) that represents the traces. In particular, the model is defined by a Deterministic Finite Automaton (DFA), where each state represents a state of the system and the transitions between states are actions taken by the user. The language of the DFA specifies the set of words that the DFA accepts. In this sense, a trace can be seen as a word, where each activity is represented by a unique letter. Therefore, a model is said to capture the behavior of a trace if and only if the corresponding DFA accepts the word that the trace represents. Formally, a solution m can be described as a tuple $(Q, \Sigma, \delta, q_0, F)$, where:

- Q is a finite set of states.
- Σ is the input alphabet, that is, the set of transitions.
- $\delta : Q \times \Sigma \rightarrow Q$ is the transition function that maps a state and an input symbol to a new state.
- $q_0 \in Q$ is the initial state.
- $F \subseteq Q$ is the set of final states.

In Fig. 1a, we present an example instance of the problem composed of the set of initial traces $W = \{w_1, w_2, w_3, w_4\}$. In Fig. 1b we represent a possible solution, m', for that instance. This solution contains 5 states ($Q = \{q_0, q_1, q_2, q_3, q_4\}$) and 5 transitions from the alphabet $\Sigma = \{$ "Open Firefox", "Load page", "Close Firefox", "Search", "Login"$\}$. In this case, we can observe that the DFA proposed as solution (m') is able to accept all traces in W.

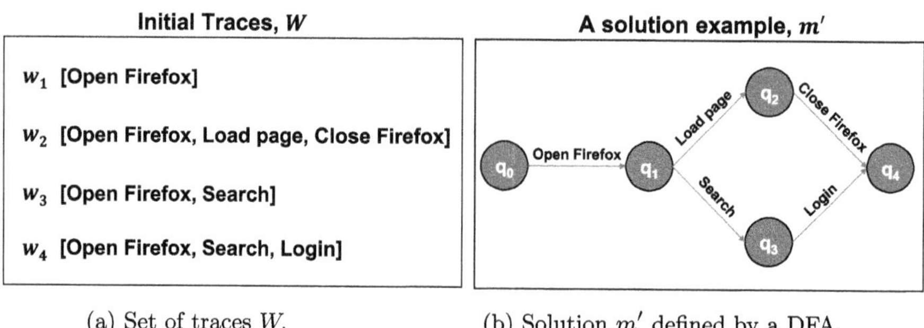

(a) Set of traces W.   (b) Solution m' defined by a DFA.

Fig. 1. Example of a problem instance, represented by the set of traces W, and an arbitrary solution m' for that instance.

To evaluate the quality of the solutions, three objective functions are considered. The first one is known as Under-Approximation (UA). This objective function counts the number of input traces that are not represented by the model (i.e., that are not accepted by the DFA). Formally, UA is defined in Eq. 1, where

m is the model, W is the input set of traces, and $L(m)$ denotes the language of the DFA (i.e., the set of words accepted by the DFA). Notice that the goal is to minimize the value of this objective function. Thus, an optimal solution that represents every trace will have a value of $\text{UA} = 0$.

$$\text{UA}(m, W) = \sum_{w \in W} \begin{cases} 0 \text{ if } w \in L(m) \\ 1 \text{ otherwise.} \end{cases} \tag{1}$$

Example. In the example of Fig. 1, the solution m' represents all the traces in W. Therefore, it achieves an optimal under-approximation value, i.e., $\text{UA} = 0$. Now, consider the case where there exists an additional trace $w_5 = [$"Open Firefox", "Settings", "Close Firefox"$]$ that is not accepted by the solution m'. In that case, m' would have $\text{UA} = 1$.

The second objective function is Over-Approximation (OA). This function measures the number of words accepted by the model that do not appear in the input set of traces. That is, the behavior represented by the model that is not part of the system. Formally, OA is defined in Eq. 2, where m is the model, W is the input set of traces, and $P(m) \subseteq L(m)$ is the set of words accepted by the model. Notice that for practical reasons, $P(m)$ only contains words from $L(m)$ with a length less than or equal to four. This limitation is defined in the literature to reduce the computational cost of the OA objective function, where considering paths of any length might be impractical for large models. The goal is again to minimize the value of the objective function. Therefore, a solution with $\text{OA} = 0$ accepts only the traces observed in W.

$$\text{OA}(m, W) = \sum_{p \in P(m)} \begin{cases} 0 \text{ if } p \in W \\ 1 \text{ otherwise.} \end{cases} \tag{2}$$

Example. In the example of Fig. 1, every sequence accepted by m' also belongs to W, which means that the solution does not introduce any behavior outside the observed traces. Therefore, it achieves an optimal over-approximation value, i.e., $\text{OA} = 0$. Now, suppose that the trace $w_4 = [$"Open Firefox", "Search", "Login"$]$ is not part of W. In that case, the sequence "Open Firefox", "Search", "Login" accepted by m' would not belong to W, resulting in $\text{OA} = 1$.

The third objective function is to minimize the size of the model (SIZE), measured as the number of states. Formally, it is defined in Eq. 3, where Q is the set of states. Smaller models are preferable as they are easier for developers to understand and use to derive meaningful test cases.

$$\text{SIZE}(m) = |Q|. \tag{3}$$

Example. The size of m' in Fig. 1 corresponds to the number of states contained in DFA, which is equal to 5. Thus, $\text{SIZE}(m) = 5$.

The goal of the optimization problem defined in this paper is to generate models that faithfully represent the activities described by the input set of traces while balancing model size and avoiding unrecognized behaviors. These objectives are inherently conflicting. The larger the set of behaviors that might be represented, usually imply a higher size of the solution. On the other hand, smaller models might be unable to represent the whole spectrum of activities. This trade-off promotes tackling this as a multi-objective optimization problem, since a compromise between different objectives needs to be considered.

Finally, the literature [8] imposes two practical constraints on the models: the value of the OA objective function must be less than 500 and the solution SIZE must be smaller than 500 states. This prevents generating overly complex and large models that are not useful in practice for software engineers.

3   Algorithmic Proposal

In this work, we propose the use of the Variable Neighborhood Search (VNS) framework. VNS, based on the systematic exploration of different neighborhood structures, was proposed by Mladenović and Hansen in 1997 [12]. Given the success of VNS-based methods in multiple single-objective optimization problems, Duarte *et al.* proposed an extension of VNS for multi-objective optimization problems denoted as Multi-Objective VNS (MO-VNS) [7].

Among the different proposals within the MO-VNS framework, in this paper we focus on the Multi-Objective Variable Neighborhood Descent (MO-VND) scheme. MO-VND systematically explores different neighborhood structures. To that aim it considers a set of non-dominated solutions as starting points for the search. For each point it explores the defined neighborhoods considering only one of the objective functions at a time. The pseudocode of MO-VND is detailed in Algorithm 1. This method receives three sets as input parameters: a set of non-dominated solutions (E), a set of neighborhood structures (N), and a set of objective functions (R). Notice, that this procedure do not consider a way to construct the initial set of solutions. In this case E is conformed with the non-dominated solutions produced using the random constructive method introduced in [8] after 1000 iterations. The first step of our MO-VND is to initialize the set of visited solutions per objective (step 2) and the variable i, which represents each objective (step 3). The method then proceeds to iteratively improve each solution x' in the set E, considering the objective i (step 5). That is, in each iteration, a non-exploited solution is selected (6), improved (step 7) and marked as visited (step 8). This ends when every solution has been exploited considering the objective i. The improvement phase is performed by a VND-i procedure, which is detailed later, and returns a set of non-dominated solutions. Steps 10-12 test if any of the solutions found by VND-i, qualifies to be added to the set E (i.e., they are not dominated). In that case, i is reset to 1. Otherwise, i is increased to consider the next objective in the following iteration (step 14). This process is repeated until every objective has been considered without finding new non-dominated solutions (step 4).

The pseudocode of the VND-i procedure is described in Algorithm 2. This method receives three input parameters: an initial solution (x), a set of neighborhood structures (N) and the current objective (i). First, the variable k, which indicates the current neighborhood being explored, is set to 1 (step 2). Additionally, the set of non-dominated solutions E is initialized with the solution x

---

Algorithm 1. Pseudocode of Multi-Objective VND (MO-VND)

1: **procedure** MO-VND(E, N, R)
2:     $S_1 \leftarrow \emptyset$, $S_2 \leftarrow \emptyset$, ..., $S_{|R|} \leftarrow \emptyset$
3:     $i \leftarrow 1$ ▷ Starts with first objective
4:     **while** $i \leq |R|$ **do**   ▷ While not all objectives have been considered
5:         **while** $E \setminus S_i \neq \emptyset$ **do**   ▷ While not all solutions have been explored
6:             $x' \leftarrow$ SELECT($E \setminus S_i$)   ▷ Random selection from non-exploited solutions
7:             $E_i \leftarrow$ VND-$i(x', N, i)$
8:             $S_i \leftarrow S_i \cup E_i$
9:         **end while**
10:        **if** $\exists\, x \in S_i : x \nprec x' \;\forall\, x' \in E$ **then**
11:            UPDATE(E, S_i)
12:            $i \leftarrow 1$
13:        **else**
14:            $i \leftarrow i + 1$
15:        **end if**
16:    **end while**
17:    **return** E ▷ Return optimized set
18: **end procedure**

(step 3). At each iteration, the algorithm explores the k-th neighborhood of the incumbent solution x ($N_k(x)$). If there exists a neighbor solution better than x (step 5), then it is added to the set (step 6), the incumbent solution is replaced by the new one (step 7), and k is reset (step 8). Otherwise, k is increased to explore the next neighborhood of x (step 10). Notice that f_i represents the evaluation of the objective function i, and the objective functions considered in this paper are described in Sect. 2. This process is repeated until every neighborhood has been explored without finding a better solution than the previous. That is, until $k > |N|$ (step 4).

Algorithm 2. Pseudocode of VND-i

1: **procedure** VND-$i(x, N, i)$
2:     $k \leftarrow 1$
3:     $E \leftarrow \{x\}$
4:     **while** $k \leq |N|$ **do**
5:         **if** $\exists\, x' \in N_k(x) : f_i(x') < f_i(x)$ **then**
6:             UPDATE(E, x')
7:             $x \leftarrow x'$
8:             $k \leftarrow 1$ ▷ Return to the initial neighborhood
9:         **else**
10:            $k \leftarrow k + 1$ ▷ Move to next neighborhood
11:        **end if**
12:    **end while**
13:    **return** E ▷ Return non-dominated solutions
14: **end procedure**

To apply the `MO-VND` method, we propose five distinct operators. The design of these operators is motivated by the structure of a DFA, which is composed of two entities: states and transitions. Consequently, to effectively explore the search space, the proposed operators are designed to systematically modify these building blocks. Specifically, they cover the addition and removal of both states and transitions, including self-loops. In Fig. 2 we illustrate the effect of each operator when applied to an initial example solution m'', depicted in Fig. 2a. All transitions depicted in the figure are assumed to belong to the alphabet $\Sigma = \{$ *"Load page"*, *"Reload"*, *"Open Firefox"*, *"Refresh page"*, *"Login"*, *"Search"*, *"Close Firefox"*$\}$. Notice that the transition *"Refresh page"* is not used in the solution depicted in Fig. 2a.

The first proposed neighborhood structure (N_1) is based on the operator denoted `AddTransition`. This operator inserts a new transition, among the existing ones in Σ, between two existing states in Q. In Fig. 2b, we depict a solution obtained by applying the `AddTransition` operator to the initial solution. In particular, a new transition labeled *"Refresh page"* is added from the state q_2 to q_5. The size of this neighborhood is equal to $|\Sigma| \cdot |Q|^2$, since any transition can be added from any state to any other state in the DFA.

The second proposed neighborhood structure (N_2) is based on the operator denoted `AddState`. This operator inserts a new state. Then it adds a new transition (among the existing ones) from an existing state to the new one. In Fig. 2c, a new state q_6 is added to the initial solution. In addition, the transition *"Refresh page"* is added from state q_2 to q_6. The size of this neighborhood is equal to $|\Sigma| \cdot |Q|$, since any transition can be added from any state in the DFA to the newly created state.

The third proposed neighborhood structure (N_3) is based on the operator denoted `RemoveState`. This operator removes a state from the DFA. After the removal, any transitions that starts or ends in the removed state are also deleted, as well as any other state that is no longer reachable from the initial one. In Fig. 2d, we show a solution obtained by removing the state q_2 from the initial solution. Notice that the transitions *"Load page"*, *"Close Firefox"* and the state q_3 have also been removed. The size of this neighborhood is equal to $|Q|$.

The fourth proposed neighborhood structure (N_4) is based on the operator denoted `AddLoop`. This operator creates a loop in a state using a new transition. In Fig. 2e, a loop with the transition *"Refresh page"* is added to state q_2. The maximum size of this neighborhood is $|\Sigma| \cdot |Q|$. In practice, this neighborhood only considers adding loops to states without loops.

The fifth proposed neighborhood structure (N_5) is based on the operator denoted `RemoveLoop`. This operator deletes a loop from a state that has one. This is only applied if the solution has more than one transition. In Fig. 2f, the loop of state q_4 is removed. The maximum size of this neighborhood is $|\Sigma| \cdot |Q|$. However, the size is usually smaller. These neighborhood structures have been designed to address the key structural components of the solutions.

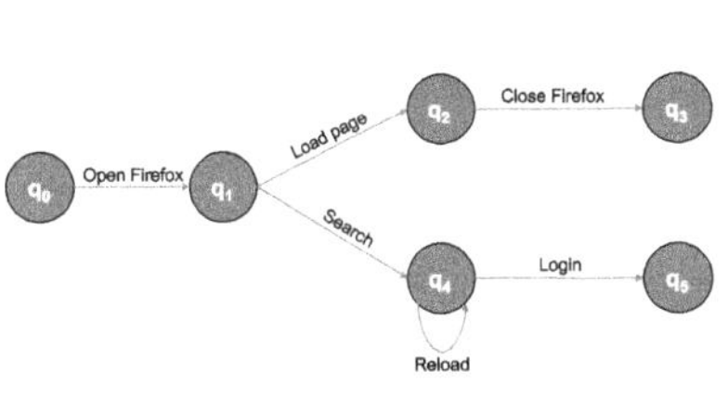

(a) Initial solution m''.

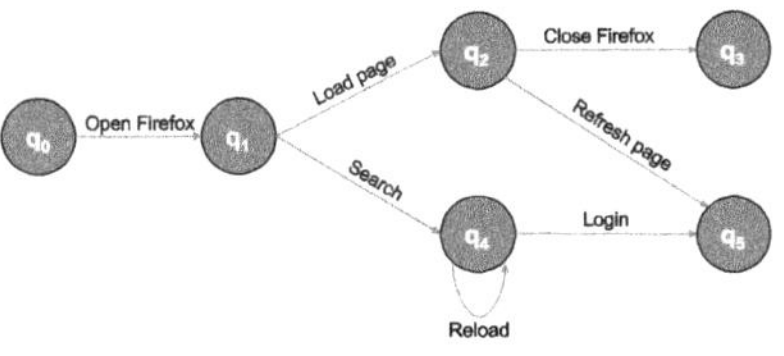

(b) Example of a solution in the neighborhood N_1 obtained by applying the operator **AddTransition** to m''.

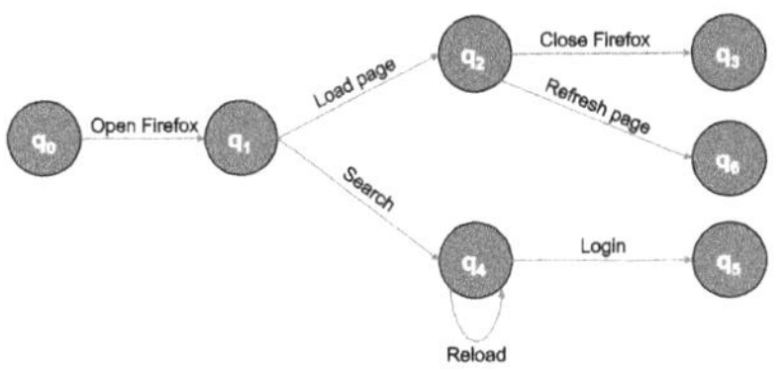

(c) Example of a solution in the neighborhood N_2 obtained by applying the operator **AddState** to m''.

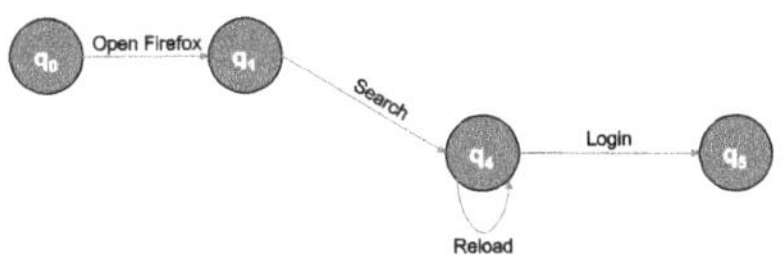

(d) Example of a solution in the neighborhood N_3 obtained by applying the operator **RemoveState** to m''.

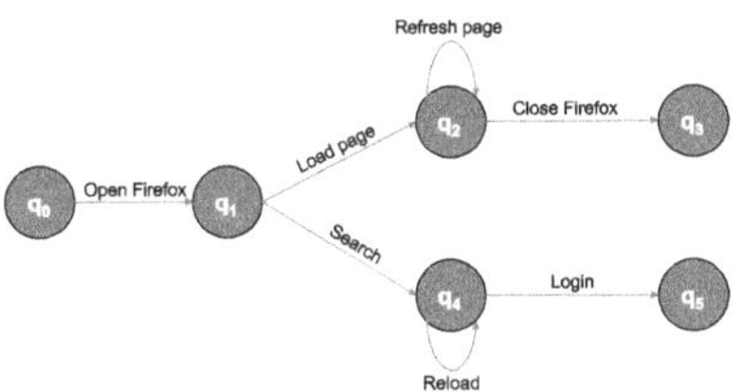

(e) Example of a solution in the neighborhood N_4 obtained by applying the operator **AddLoop** to m''.

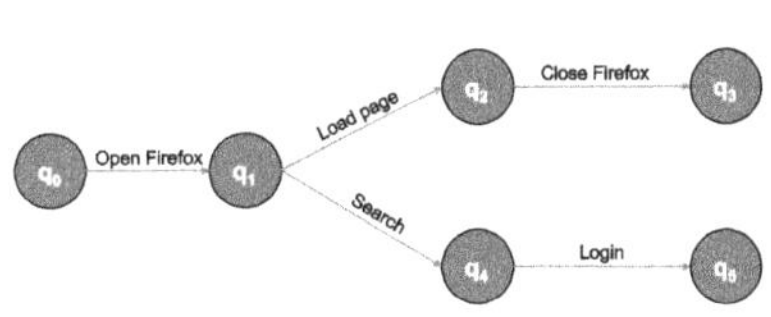

(f) Example of a solution in the neighborhood N_5 obtained by applying the operator **RemoveLoop** to m''.

Fig. 2. An initial solution and a set of solutions for each neighborhood based on different operators.

4   Experimental Results

This section presents the main experimental results obtained during the development of this work. To evaluate the proposed algorithms, we used a set of benchmark instances made publicly available by Guizzo *et al.* [8]. These instances were derived from real-world software systems by processing issue tracking platforms such as Bugzilla[1]. In total, the benchmark contains 10 different instances, each associated with a distinct software system. For a more detailed analysis of the characteristics of the instances, we refer the reader to Appendix A.

All experiments were conducted on a server with an AMD EPYC 7282 processor (8 cores), 32GB of RAM, and Ubuntu 20.04.1 LTS. All methods were

---

[1] https://www.bugzilla.org/.

implemented in Python 3.12 and run with a time limit of 60 min. In every experiment, the initial set of solutions for the MO-VND algorithm is generated by the constructive procedure proposed by Guizzo *et al.* [8]. This method randomly selects some traces from the given instance, builds models to represent them, and then merges those models. By using the same constructive procedure, we ensure a fair comparison with the state of the art.

4.1   Quality Indicators

When assessing solution quality in multi-objective optimization problems, the literature generally recommends using appropriate Quality Indicators (QIs) instead of evaluating each objective independently [1]. Nonetheless, choosing the most suitable QI is a complex task and lacks a general agreement within the realm of SBSE [11].

In general, QIs can capture four different quality aspects: convergence, spread, uniformity, and cardinality. However, no single QI covers all these aspects simultaneously. For this reason, and following recommendations from the literature [1,11,16], we selected three QIs that collectively evaluate all desired quality aspects: Hypervolume (HV), Inverted Generational Distance Plus (IGD+), and Pareto Front Size (PFS).

The HV indicator evaluates convergence, spread, and uniformity. Higher HV values indicate better overall quality. The IGD+ metric measures the proximity of the evaluated set to a reference front. Lower IGD+ values correspond to higher quality. Finally, the PFS represents the number of non-dominated solutions in the set. Usually, a large number is preferred.

The IGD+ QI requires a reference front. Although this front ideally corresponds to the optimal Pareto front, such a reference is rarely available in practice. Following established literature [1], an approximate reference front is constructed for each experiment by aggregating all the non-dominated solutions produced by the compared methods.

4.2   Comparison of Neighborhood Structures

To evaluate the effectiveness of each neighborhood structure in isolation, we performed a preliminary experiment. In this experiment, we run the proposed method once per neighborhood structure. In each run, only one of the neighborhoods was explored. Notice that we did not include neighborhood N_5 (based on removal of loops) in this experiment, since the constructive method does not generate loops in the initial solutions [8]. As a baseline, we include the results obtained by the constructive method.

In Table 1, we show the results obtained by the methods being compared. Particularly, we consider just the constructive (as a baseline) and its combination with a local search (LS) that only explores one of the neighborhoods in isolation. For each row, we show the average PFS, HV, and IGD+ values of the Pareto fronts obtained by the method. CPUt (s) shows the average computing time and (# Iter./s) shows the number of iterations explored per second (not considering the time spent by the constructive procedure). Rows are sorted by descending HV value and the best result for each QI is highlighted in bold font. The results in Table 1 show that the method which explores neighborhood structure N_2 is the most effective one. Exploring this neighborhood leads to the achievement of the best values for every QI, indicating better convergence and diversity than the other methods. However, it is also the most computationally expensive neighborhood, although this might be partially explained by its ability to improve the initial solutions. On the other hand, exploring the neighborhood N_3 leads to obtaining low-quality Pareto fronts. Moreover, note that exploring this neighborhood results in the fastest method of the comparison. Again, this might be partially explained by its inability to improve the initial solutions. Nonetheless, it is also true that N_3 is the smallest neighborhood structure in the comparison. In contrast, exploring either N_4 or N_1 leads to moderate performance, balancing quality and computing time. Finally, note that every method evaluated led to improving the initial solutions generated by the constructive procedure.

Table 1. Comparison of the performance of the proposed method when exploring each neighborhood structure in isolation.

Method	PFS	HV	IGD+	# Iter./s	CPUt (s)
Constructive + LS (N_2/`AddState`)	**57.20**	**0.5350**	**0.0967**	**334.03**	2,560.26
Constructive + LS (N_4/`AddLoop`)	24.50	0.3517	0.3178	111.35	862.87
Constructive + LS (N_1/`AddTransition`)	16.00	0.3444	0.3422	68.13	1,176.02
Constructive + LS (N_3/`RemoveState`)	9.90	0.2713	0.3854	286.84	25.89
Constructive	6.40	0.2652	0.3983	–	25.51

4.3   Comparison of the Order of Neighborhood Structures

As it is commonly known in the literature, the ordering in which neighborhoods are explored within VNS methods is crucial for their performance. In this experiment we test different neighborhood orderings. In particular, two configurations were evaluated. The first one explores the neighborhood structures in descending order considering the HV value obtained in the previous experiment (see Sect. 4.2): N_2, N_4, N_1, N_3, and N_5. The second one explored the neighborhood structures in ascending order considering the computational time consumed in the previous experiment (see Sect. 4.2): N_3, N_4, N_1, N_2, and N_5. Note that we have included N_5 (which was not evaluated in the previous experiment) in the last position.

In Table 2, we summarize the results obtained. We include the average PFS, HV, and IGD+ values of the Pareto fronts obtained by each method, as well as the average computing time. In addition, we include the performance of the method when exploring each neighborhood structure in isolation and the set of initial solutions obtained by the constructive procedure. Rows are sorted by descending HV value and the best result for each QI is highlighted in bold font. As it can be seen, the first configuration demonstrates superior performance for every QI. In contrast, the second configuration shows worse performance than exploring only the neighborhood N_2 and is even more computationally expensive.

Table 2. Comparison of the performance of the proposed method when exploring neighborhood structures in different order.

Method	PFS	HV	IGD+	CPUt (s)
MO-VND(N_2, N_4, N_1, N_3, N_5)	**61.70**	**0.5158**	**0.1319**	3,600.00
Constructive + LS (N_2/AddState)	54.90	0.4660	0.1934	2,560.26
MO-VND(N_3, N_4, N_1, N_2, N_5)	51.70	0.4609	0.1939	3,600.00
Constructive + LS (N_4/AddLoop)	24.50	0.2984	0.3721	862.87
Constructive + LS (N_1/AddTransition)	16.00	0.2846	0.4002	1,176.02
Constructive + LS (N_3/RemoveState)	9.90	0.2149	0.4589	25.89
Constructive	6.40	0.2112	0.4631	25.51

These findings demonstrate that performance is heavily influenced by the order of neighborhoods. Moreover, an interesting takeaway is that some orderings may even lead to worse results than only exploring one of the neighborhood structures. Interestingly, one common way of ordering neighborhoods is from least to most computationally expensive, which in this case led to bad performance. Therefore, there is no clear criterion for ordering neighborhoods other than experimentally evaluating all possible orderings. However, given the high number of possible combinations (120 in total), we have not tried all. Instead, we evaluated two orderings given the information at hand. That is, the quality and computational time obtained by each configuration investigated in Sect. 4.2. Given the results obtained, we decided to configure the proposed method to explore neighborhood structures in the following order: N_2, N_4, N_1, N_3, and N_5.

4.4   Comparison with the State-of-the-Art Methods

This section compares the performance of the proposed MO-VND algorithm with the state-of-the-art methods proposed in the literature. In particular, we include the methods proposed by Guizzo *et al.* [8], which are based on: NSGA-II [5], NSGA-III [4], and MOEA/D [17]. For these methods, the authors introduced several domain-specific crossover and mutation operators. Due to space constraints, a detailed explanation of these techniques is omitted, and readers are referred to the original publications [4,5,8,17]. We have used the original implementations made publicly available by the authors[2] [8]. Note that, as previously

---

[2] https://github.com/SOLAR-group/ModelInference

described in Sect. 4, all experiments are run in the same computing environment and coded with the same programming language. Moreover, all methods, including MO-VND, use the same constructive procedure.

In Table 3, we present the results obtained. For each method, we report the average value obtained for each QI and the average computing time consumed (CPUt (s)). Rows are sorted by descending HV value. The best result for each QI is highlighted in bold font. As can be seen, NSGA-II outperforms the other methods for every QI evaluated. MOEA/D ranks second, showing competitive values, while NSGA-III performs significantly worse in comparison. In contrast, the proposed MO-VND approach does not match the quality of the solutions found by the evolutionary methods, with lower scores in HV and IGD+.

Table 3. Comparison of the performance of the proposed procedure versus the state-of-the-art methods.

Method	PFS	HV	IGD+	CPUt (s)
NSGA-II	**332.20**	**0.6603**	**0.0101**	3,600.00
MOEA/D	112.90	0.4978	0.0828	3,600.00
NSGA-III	30.90	0.3924	0.1386	3,600.00
MO-VND	61.70	0.2500	0.3093	3,600.00

In Fig. 3 and Fig. 4, we illustrate the HV and IGD+ obtained by the methods over time. In particular, we depict the average values obtained by the Pareto fronts generated by each method at every minute during the search process. As

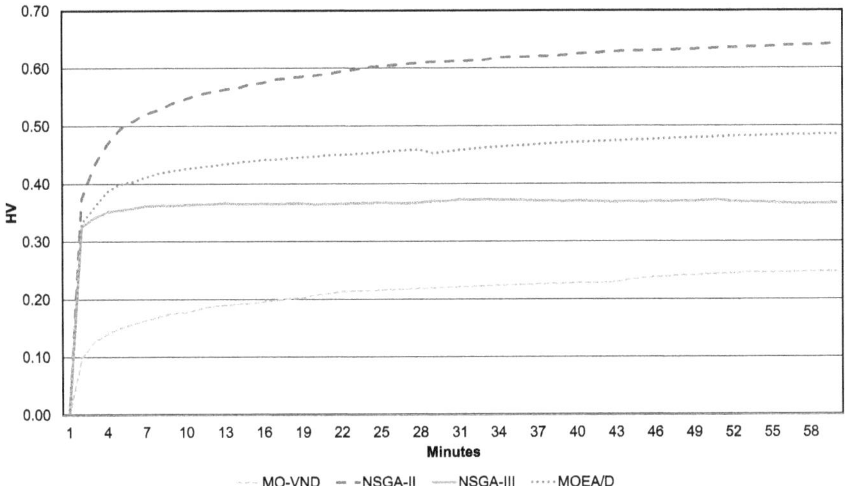

Fig. 3. Average HV value of the solutions found by each method during the search process.

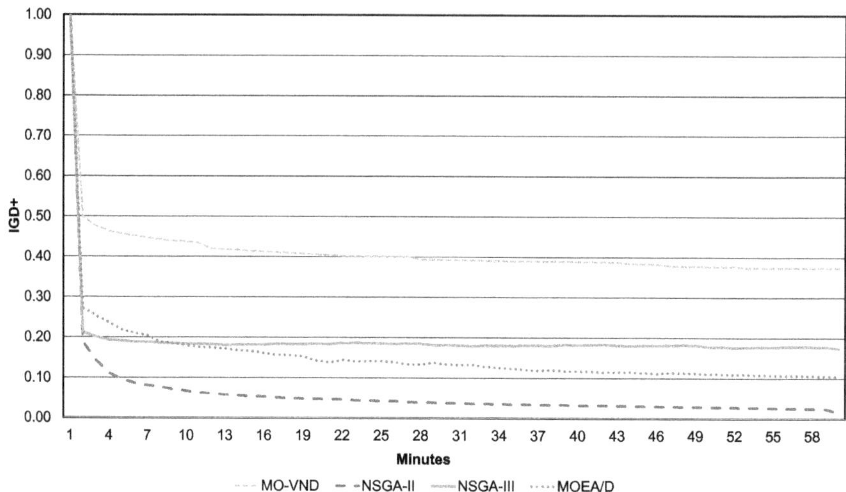

Fig. 4. Average IGD+ value of the solutions found by each method during the search process.

expected, NSGA-II exhibits the best performance in both indicators. Moreover, this better performance is shown from the first minutes. In contrast, MOEA/D shows a worse performance. Interestingly, NSGA-III seems to become stagnated, showing little to no progress from the first 5 min to the end of the process. The proposed MO-VND method still shows progress at the end of the time horizon.

5   Conclusions and Future Work

In this paper, we studied the multi-objective optimization problem of model inference from user bug reports. For this problem, we proposed a method based on the MO-VNS framework. In particular, the proposed method was based on the MO-VND scheme. Moreover, we proposed five different neighborhood structures based on specific knowledge of the problem. Then, we evaluated the performance of the method when exploring the different neighborhoods in isolation and also when exploring all neighborhoods with different orderings. As is commonly known in the literature, the order in which neighborhoods are explored is important for the performance of VNS-based methods. Interestingly, however, the results showed that some orderings achieve worse quality than exploring some of the neighborhoods in isolation. At least during the time limit of the experiment.

Finally, we compared the performance of the proposed method with the state-of-the-art algorithms proposed in the literature. The results showed that the proposed MO-VND method is promising, but still does not achieve the same performance than the other methods included in the comparison. However, there is still room for improvement.

In future work, it would be interesting to study strategies to efficiently evaluate the objective functions and identify promising solutions within each neighborhood structure. These strategies would greatly improve the efficiency of the method. In addition, given the importance of the order of the neighborhoods highlighted in this work, it would be interesting to analyze further combinations, using tools for the automatic configuration of the algorithm. Finally, since MO-VND is based on local search procedures, its performance may be strongly influenced by the initial solutions. Designing a better constructive method could improve the performance of the method. Moreover, introducing a shaking phase could help the method become less dependent on the randomly generated initial solution.

Acknowledgements. This research has been partially supported by grants: PID2021-125709OA-C22 and PID2024-160226OB-C22, funded by MCIN/AEI/ 10.13039/501100011033 and "ERDF A way of making Europe"; CIRMA-CM Ref. TEC-2024/COM-404 funded by Comunidad Autónoma de Madrid; TSI-100930-2023-3 (MCA07) funded by Ministerio para la Transformación Digital y de la Función Pública; "Cátedra de Innovación y Digitalización Empresarial entre Universidad Rey Juan Carlos y Second Episode" (Ref. ID MCA06); RED2022-134480-T (Ref. Int: M3167) funded by Agencia Estatal de Investigación; and 2024/SOLCON-135988, and 2024/SOLCON-64631, funded by Universidad Rey Juan Carlos.

Appendix A Detailed description of the dataset

In Table 4, we present some statistics of the instances in the dataset. In particular, we include the number of traces and the maximum, minimum, mean, and standard deviation of the number of activities per trace for each instance. Rows are sorted by the number of traces in ascending order. As can be observed, instances are of varying sizes, ranging from 501 to 5,451 traces. The size of the traces within instances are also of varying sizes, ranging from 1 to 77 transitions. Each instance contains an average of 1894.20 traces. On average, the number of transitions per trace is 5.78, with a standard deviation of 3.44. The largest trace per instance contains an average of 43.2 transitions or activities, and the shortest contains an average of 1.3.

Table 4. Number of traces and maximum, minimum, mean, and standard deviation of the number of activities per trace for each instance in the dataset.

Instance	# Traces	Number of transitions per trace			
		Max.	Min.	Mean	Std. Dev.
Kate	501.00	25.00	1.00	5.17	3.01
Vibe	531.00	77.00	2.00	8.23	5.74
Krita	1,202.00	57.00	1.00	5.46	3.68
LibreOffice	1,300.00	72.00	1.00	5.97	4.51
Firefox_for_Android	1,318.00	29.00	2.00	6.04	3.63
Firefox_OS	1,353.00	39.00	2.00	6.00	3.17
SeaMonkey	1,998.00	20.00	1.00	4.56	2.31
Thunderbird	2,068.00	31.00	1.00	5.17	3.26
Calendar	3,220.00	53.00	1.00	5.42	3.26
BIRT	5,451.00	29.00	1.00	5.81	3.15
Average	1,894.20	43.20	1.30	5.78	3.44

Note: All values have been rounded to two decimal places.

References

1. Ali, S., Arcaini, P., Pradhan, D., Safdar, S.A., Yue, T.: Quality indicators in search-based software engineering: An empirical evaluation. ACM Trans. Softw. Eng. Methodol. (TOSEM) **29**(2), 1–29 (2020)
2. Boussaïd, I., Siarry, P., Ahmed-Nacer, M.: A survey on search-based model-driven engineering. Autom. Softw. Eng. **24**(2), 233–294 (2017)
3. Chen, C., Alfayez, R., Srisopha, K., Boehm, B., Shi, L.: Why is it important to measure maintainability and what are the best ways to do it? In: 2017 IEEE/ACM 39th International Conference on Software Engineering Companion (ICSE-C), pp. 377–378. IEEE (2017)
4. Deb, K., Jain, H.: An evolutionary many-objective optimization algorithm using reference-point-based nondominated sorting approach, part I: solving problems with box constraints. IEEE Trans. Evol. Comput. **18**(4), 577–601 (2013)
5. Deb, K., Pratap, A., Agarwal, S., Meyarivan, T.A.M.T.: A fast and elitist multi-objective genetic algorithm: NSGA-II. IEEE Trans. Evol. Comput. **6**(2), 182–197 (2002)
6. Dennis, A., Wixom, B., Tegarden, D.: Systems Analysis and Design : An Object-Oriented Approach with UML. Wiley (2015)
7. Duarte, A., Pantrigo, J.J., Pardo, E.G., Mladenovic, N.: Multi-objective variable neighborhood search: an application to combinatorial optimization problems. J. Global Optim. **63**(3), 515–536 (2014)
8. Guizzo, G., Califano, F., Sarro, F., Ferrucci, F., Harman, M.: Inferring test models from user bug reports using multi-objective search. Empir. Softw. Eng. **28**(4), 95 (2023)
9. Harman, M., Jones, B.F.: Search-based software engineering. Inf. Softw. Technol. **43**(14), 833–839 (2001)

10. International Organization for Standardization: ISO/IEC/IEEE 12207:2017 Systems and software engineering — Software life cycle processes. International Organization for Standardization (ISO), International Electrotechnical Commission (IEC), Institute of Electrical and Electronics Engineers (IEEE) (2017)
11. Li, M., Chen, T., Yao, X.: How to evaluate solutions in pareto-based search-based software engineering: a critical review and methodological guidance. IEEE Trans. Softw. Eng. **48**(5), 1771–1799 (2020)
12. Mladenović, N., Hansen, P.: Variable neighborhood search. Comput. Oper. Res. **24**(11), 1097–1100 (1997)
13. Molnar, A.-J., Motogna, S.: A study of maintainability in evolving open-source software. In: Ali, R., Kaindl, H., Maciaszek, L.A. (eds.) ENASE 2020. CCIS, vol. 1375, pp. 261–282. Springer, Cham (2021)
14. Wong, W.Y., Yu, S.W., Too, C.W.: A systematic approach to software quality assurance: the relationship of project activities within project life cycle and system development life cycle. In: 2018 IEEE Conference on Systems, Process and Control (ICSPC), pp. 123–128. IEEE (2018)
15. Yin, Z., Yuan, D., Zhou, Y., Pasupathy, S., Bairavasundaram, L.: How do fixes become bugs? In: Proceedings of the 19th ACM SIGSOFT Symposium and the 13th European Conference on Foundations of Software Engineering, pp. 26–36 (2011)
16. Yuste, J., Pardo, E.G., Duarte, A., Hao, J.K.: Multi-objective general variable neighborhood search for software maintainability optimization. Eng. Appl. Artif. Intelli. **133**, 108593 (2024)
17. Zhang, Q., Li, H.: MOEA/D: a multiobjective evolutionary algorithm based on decomposition. IEEE Trans. Evol. Comput. **11**(6), 712–731 (2007)
18. Zhang, Y., Harman, M., Jia, Y., Sarro, F.: Inferring test models from Kate's bug reports using multi-objective search. In: Barros, M., Labiche, Y. (eds.) SSBSE 2015. LNCS, vol. 9275, pp. 301–307. Springer, Cham (2015)

Heuristics for the PVRP with Driver Consistency: An Applied Case Study

Silvia Ventura-Cabrejas$^{(\boxtimes)}$, Sergio Pérez-Peló, and Jesús Sánchez-Oro

Department of Computer Science and Statistics, Universidad Rey Juan Carlos,
Madrid, Spain
`{silvia.ventura,sergio.perez.pelo,jesus.sanchezoro}@urjc.es`

Abstract. This work focuses on implementing a metaheuristic algorithm for solving the Periodic Capacitated Vehicle Routing Problem with Time Windows and Driver Consistency (PCVRPTW-DC), a problem that hybridizes the Consistent Vehicle Routing Problem and the Vehicle Routing Problem with Time Windows. Additionaly, the final goal of the work is the development of a customized solution for weekly logistic route planning in a distribution company. Initially, the planning was manually generated by company experts and, later, by using a commercial solver (Hexaly). Due to new flexibility and technological independence requirements, it was decided to replace this tool with an in house system. Experimental results show that this new system provides competitive and even superior solutions compared with the previous approaches, establishing the groundwork for a more adaptable and efficient solution.

Keywords: Metaheuristics · Heuristics · GRASP · ALNS · VND

1 Introduction

In logistics operations, the "last mile", referring to the final segment from logistics centers to final customers, requires special attention. This stage is notably complex due to several factors: restricted vehicle capacity, strict delivery time windows, high demand variability, and widespread geographic spread. Consequently, optimizing this stage is crucial to managing operational costs effectively while ensuring consistency in service quality.

The study and optimization of delivery routes has a long history, beginning with the Traveling Salesman Problem (TSP), formalized in [1] while seeking efficient school bus routes. This foundational work inspired a broad family of problems, including the Capacitated Vehicle Routing Problem (CVRP) introduced in [2], and the Capacitated Vehicle Routing Problem with Time Windows (CVRPTW), first proposed in [3]. A further extension, the Periodic Capacitated Vehicle Routing Problem (PCVRP) [4], accounts for multiple visits over a planning horizon. Within this periodic setting, the concept of driver consistency [5], ensures that frequently visited customers are served by the same driver,

S. Cavero et al. (Eds.): ICVNS 2025, LNCS 16256, pp. 196–210, 2026.
https://doi.org/10.1007/978-3-032-19582-1_14

improving service quality, strengthening customer relationships, and enhancing operational efficiency.

This work addresses the Periodic Capacitated Vehicle Routing Problem with Time Windows and Driver Consistency (hereafter, PCVRPTW-DC). The problem arises in a distribution company[1] that, until relatively recently, managed its weekly route planning manually. These decisions relied heavily on the expertise of a small group of employees, creating risks to efficiency and continuity, particularly in cases of staff turnover or absences.

To professionalize and automate the process, the company initially adopted a commercial solver, which enabled the automatic generation of high quality solutions. Over more than a year, the configuration of the solver and internal structure were carefully adapted to the specific needs of the company, delivering satisfactory operational performance.

However, new requirements and limitations soon emerged from the use of an external tool. These included technological dependency, limited flexibility for modifications, and the need for stronger integration with internal processes. These factors ultimately motivated a migration toward proprietary software, providing more accessibility and control without compromising solution quality.

This study focuses precisely on that transition: from an approach based on a commercial solver to a custom developed system. The objective is to demonstrate that it is possible to replicate, and even improve, the performance achieved with commercial tools through a customized implementation fully adapted to the logistics problem at hand. By developing a functional prototype and conducting experimental validation with real world data, this work analyzes the feasibility and potential of the proposed solution.

1.1   Problem Definition

The PCVRPTW-DC consists of planning daily delivery routes for a fleet of vehicles over a periodic horizon, such that each customer is served according to their required frequency and within specific time windows. At the same time, it is required that each customer is always visited by the same driver for all their deliveries.

The mathematical formulation adopted in this work builds upon the model proposed by [6] for the PCVRP with Driver consistency, while incorporating the time window constraints following the classical formulation of [7]. This combined framework provides the basis for addressing the problem under study. More specifically, the objective of the proposal is to hierarchically optimize these objective functions:

$$\min \quad \sum_{t \in T} \sum_{k \in K} y_0^{t,k} \tag{1}$$

$$\min \quad \sum_{t \in T} \sum_{k \in K} r_k^t \tag{2}$$

---

[1] Due to privacy reasons, no specific information about the company or its clients is provided.

The first objective (1) minimizes the total number of routes executed over the planning horizon, and the second (2) minimizes the total driving time of all vehicles. However, since this is a real-world application, additional restrictions beyond the classical formulations are required due to the business-specific nature of the problem. The following constraints are considered:

- Routes are planned from Monday to Saturday.
- A maximum heterogeneous fleet is available and can be assigned every day.
- While the full fleet is available from Monday to Friday, only half of it is used on Saturdays. This specific feature makes the use of templates non-viable, so a different strategy must be followed for solving these days.
- There is a subset of priority customers who cannot share a route with more than a limited number of other priority customers. This is not a direct problem constraint, but a client-imposed one.
- Priority customers cannot be reassigned to different vehicles.
- Visit patterns (where a pattern is a subset of days indicating on which days a customer must be served) must be spaced throughout the week.
- Some customers cannot be visited on Saturdays, so their assigned visit pattern must not include this day.
- Regarding service consistency, three constraints are imposed: the overall service consistency for all customers must be at least equal to a percentage established by the company; the same constraint applies to the subset of priority customers; and a minimum percentage of priority customers must achieve perfect service consistency.

Service consistency is calculated as the percentage of deliveries made by the assigned driver (the one who performs the majority of deliveries for that customer during the week). Saturdays are not considered in the consistency calculation, as they are treated as a special day.

1.2　State of the Art

To the best of our knowledge, there is no dedicated survey nor a method that explicitly addresses the PCVRPTW-DC in its entirety. However, relevant progress has been made through related problems such as the Periodic Capacitated Vehicle Routing Problem with Driver Consistency (PCVRP-DC), or Consistent Vehicle Routing Problem (ConVRP). These studies provide valuable insights through both exact and heuristic approaches, which can be considered as building blocks for tackling the PCVRPTW-DC.

In [8] the first exact method for the ConVRP is proposed. Their formulation relies on column generation, but instead of daily route-based variables, each column represents the set of routes assigned to a vehicle across the entire horizon, producing a stronger relaxation. This enables the solution of medium-sized instances, while a complementary large neighborhood search is used both as an upper bounding procedure and as a stand-alone heuristic. In [9] the PVRP-DC is studied and extended by incorporating service time optimization. They proposed a mixed-integer linear formulation and developed several branch-and-cut algorithms, including Benders-based variants. Their results highlight the effectiveness of decomposition strategies for ensuring consistency across multiple days while also optimizing service schedules. In [10] a multi-period dial-a-ride problem with driver consistency is investigated. They proposed formulations solved with branch-and-cut and complemented by a large neighborhood search. Their experiments showed that enforcing strict consistency can significantly increase costs, although the magnitude of the impact is highly instance-dependent.

Heuristic and metaheuristic approaches have been widely applied to consistency-oriented routing problems. In [11] the ConVRP with a heterogeneous fleet is analyzed, proposing a hierarchical tabu search framework enhanced with variable neighborhood descent. This method proved effective for large-scale instances and offered managerial insights into the trade-off between consistency requirements and vehicle-customer compatibility constraints. In [12] a variable neighborhood search for the ConVRP is developed, explicitly addressing consistency in both driver assignment and arrival times. Their algorithm includes tailored shaking procedures and mechanisms to handle time-difference excess across days. Experiments on benchmark instances showed that it outperforms existing approaches, positioning variable neighborhood search as a competitive framework for multi-period consistency.

Earlier work also considered periodic routing under constructive metaheuristics. In [13] authors applied scatter search to a periodic raw material pick-up problem, using a two-phase approach that first assigned orders to calendars and then constructed daily routes. This highlights the potential of population-based methods for capturing the calendar dimension in periodic routing.

Finally, in [5], the ConVRP is introduced, and constructive and improvement heuristics are proposed. In [14], the problem is generalized and solved by applying a flexible large neighborhood search. Their framework combined destroy and repair operators with local search to enhance arrival time consistency, outperforming template based methods and providing insights into the trade-off between cost and service stability.

2   Algorithmic Description

This section details the implementation used to solve the problem through metaheuristic algorithms.

Although the algorithm is organized in two phases, the Variable Neighborhood Descent (VND) [15] is the centerpiece throughout and the main source of solution quality. The first phase uses a Greedy Randomized Adaptive Search Procedure (GRASP) [16] only to diversify and generate starting points that will be immediately refined by VND. The second phase applies an Adaptive Large Neighbourhood Search (ALNS) [17], again with systematic VND improvement of every accepted solution. In short, GRASP and ALNS provide exploration, while VND is the engine that yields high-quality solutions. It is important to remark that the neighborhoods explored in the VND phase of the algorithm are traversed following a first improvement strategy. This is because in real-life scenarios, computing time is a hard constraint, and exploring the whole neighborhood and making the best move found implies an overload in this metric.

GRASP is composed of two phases as illustrated in Algorithm 1: the first generates diverse initial solutions (step 3), and the second refines each of them through VND (step 4). The best VND-refined solution over all starts is kept (step 9).

Algorithm 1. $GRASP(max\_iter, LS, \alpha)$

1: $S_{best} \leftarrow \varnothing$
2: **for** $k = 1, ..., max\_iter$ **do**
3:     $S \leftarrow Constructive(\alpha)$
4:     $S' \leftarrow VND(S, LS)$
5:     **if** $f(S') < f(S_{best})$ **then**
6:         $S_{best} \leftarrow S'$
7:     **end if**
8: **end for**
9: **return** S_{best}

The constructive routine follows Algorithm 2. First, weekly templates are created for periodic customers (customers visited more than once per week)(step 1) and assigned to a single vehicle to ensure consistency. Because these templates are only baselines, they are immediately refined with VND (step 2). Next, the planning horizon is instantiated day by day using the improved templates (step 3), and finally non-periodic customers are inserted (step 4).

Algorithm 2. $Constructive(\alpha)$

1: $template \leftarrow$ TemplateConstruction(α)
2: $template \leftarrow VND(template, LS)$
3: $S \leftarrow$ InitializePeriodic$(template)$
4: $S' \leftarrow$ InsertNonPeriodic(S)
5: **return** S'

Weekly templates are first built to assign each periodic customer to a single vehicle, ensuring full service consistency. This process is detailed in Algorithm 3. Starting from an empty template (step 1), a candidate list of customers is maintained and customers are sequentially inserted. To begin, one candidate is randomly selected and inserted into the first available vehicle, adding it to the template (steps 3-5). While there are still candidates to be assigned, for each of them the insertion cost is computed, defined as the increase in total driving time when assigning the customer to its best position, i.e., the vehicle and position within all routes that results in the minimum increment. When feasible insertions exist, a Restricted Candidate List (RCL) (steps 7-11) is formed using only those candidates that do not induce lateness with respect to time windows.

A customer is then randomly selected from the RCL and inserted in its best position (steps 12, 13 and 14). If all possible insertions would result in infeasibility, a new vehicle is opened if available (step 17). Otherwise, the customer with the smallest induced lateness is chosen and inserted (steps 19-21).

Algorithm 3. *TemplateConstruction(α)*

1: $T \leftarrow \varnothing$
2: $CL \leftarrow \{V\}$ ▷ All candidates are periodic customers
3: $v \leftarrow RND(CL)$
4: $T \leftarrow T \cup \{v\}$
5: $CL \leftarrow CL \backslash \{v\}$
6: **while** $CL \neq \varnothing$ **do**
7:      $g_{\min} \leftarrow \min_{c \in CL} g(c)$ ▷ Minimum feasible insertion cost
8:      $g_{\max} \leftarrow \max_{c \in CL} g(c)$ ▷ Maximum feasible insertion cost
9:      **if** $g_{min} \neq \infty$ and $g_{max} \neq -1$ **then**   ▷ If there are feasible insertions
10:          $\mu \leftarrow g_{\min} + \alpha(g_{\max} - g_{\min})$
11:          $RCL \leftarrow \{c \in CL : g(c) \leq \mu\}$
12:          $v \leftarrow RND(RCL)$
13:          $T \leftarrow T \cup \{v\}$
14:          $CL \leftarrow CL \backslash \{v\}$
15:      **else**
16:          **if** *areAvailableVehicles* **then**
17:              *OpenNewVehicle*
18:          **else**
19:              **if** *lateness(v)* **then**
20:                  $v \leftarrow \min_{c \in CL} lateness(c)$
21:              **end if**
22:          **end if**
23:      **end if**
24: **end while**

Because these templates are not high quality on their own, VND is applied immediately after the construction of the template as described in Algorithm 4. VND explores a sequence of neighborhoods, restarting from the first whenever an improvement is found, which avoids premature convergence [18]. The move operators that define the explored neighborhoods are ordered from lower to higher complexity: 2-Opt (within-route edge reversal), Swap (exchange two customers within or across routes), and Insertion (relocate one customer within or across routes).

Once templates are defined, a partial weekly solution is constructed using periodic customers under the following prioritized objectives:

1. Minimize truck overload.
2. Minimize weekly delays (time-window violations and excess route duration).
3. Minimize the number of weekly routes.
4. Minimize total driving time.

Algorithm 4. *VND* (S, LS)

1: $k = 1$
2: **while** $k \leq |LS|$ **do**
3:     $S' \leftarrow LS_k(S)$
4:     **if** $f(S') < f(S)$ **then**
5:         $S \leftarrow S'$
6:         $k \leftarrow 1$
7:     **else**
8:         $k \leftarrow k + 1$
9:     **end if**
10: **end while**

The instantiation of daily solutions is carried out as shown in Algorithm 5. Predefined visit patterns for periodic customers are used, balanced over the week. Given selected customers for a day, routes follow the template's vehicle and visit order. With those two components, a partial initial solution can be built. This solution follows the predefined patterns and the template's scheduling (steps 2 and 11). Each periodic customer is iteratively evaluated with all its possible patterns: the current pattern is tentatively replaced (step 10), the solution is assessed (step 11), and the change is kept only if it improves the objectives (steps 12-16). The process stops by iteration or time limits, providing a partial solution. Saturday is treated specially, since at most half the trucks are available. Its routes are solved as a CVRPTW by a greedy heuristic rather than by templates.

The insertion of non-periodic customers completes the construction with a purely greedy procedure as shown in Algorithm 6: for each iteration, all insertions across trucks and days are evaluated for every remaining candidate, the customer with the best insertion is selected (step 3), and the insertion is committed (step 4). This continues until all non-periodic customers are placed and the initial solution is completed.

After constructing the initial solution, the procedure enters an improvement phase, where VND is applied again, being the main driver of quality.

Same-day and cross-day moves are distinguished, and the operator order is maintained from lower to higher complexity: 2-Opt, Swap (within/across routes or days), Insertion (within/across routes or days, while periodic customers keep service consistency), Delay-reduction (reassign non-periodic customers with lateness to another day when feasible), and Pattern reassignment (change the visit pattern of periodic customers and, if feasible, their truck).

Each GRASP iteration differs due to randomness in template construction, and the best VND improved solution across starts is retained. This best solution is the sole input to the final ALNS phase.

ALNS extends large-neighborhood search via adaptive operator selection and a permissive acceptance rule as shown in Algorithm 7. A roulette of destroy and repair operators is maintained, whose weights are updated by performance. At each iteration, a fraction of the solution is destroyed (step 6) only among non-periodic customers and then repaired (step 7). Crucially, whenever a new solu-

---

Algorithm 5. *InitializePeriodic(template)*

1: $patterns \leftarrow predefinedPatterns()$
2: $S_{best} \leftarrow BuildFromTemplate(template, patterns)$
3: $iter \leftarrow 0$
4: $improved \leftarrow true$
5: **while** $improved$ **and** $iter < max\_iter$ **do**
6:     $improved \leftarrow false$
7:     **for all** $c \in V$ **do**
8:         **for all** $p \in P_c$ **do**
9:             $t \leftarrow CurrentPattern(p)$
10:             $new\_pattern \leftarrow ChangePattern(patterns, t)$
11:             $S' \leftarrow BuildFromTemplate(template, new\_pattern)$
12:             **if** $f(S') < f(S_{best})$ **then**
13:                 $S_{best} \leftarrow S'$
14:                 $patterns \leftarrow new\_patterns$
15:                 $improved \leftarrow true$
16:             **end if**
17:         **end for**
18:     **end for**
19:     $iter \leftarrow iter + 1$
20: **end while**
21: **return** S_{best}

Algorithm 6. *GreedyInsertion(S)*

1: CL $\leftarrow \mathcal{N}$ ▷ All candidates are non-periodic customers
2: **while** CL $\neq \varnothing$ **do**
3:     $v \leftarrow \arg\min_{c \in CL} g(c)$
4:     $S \leftarrow S \cup v$
5:     CL $\leftarrow$ CL $\setminus \{v\}$
6: **end while**

---

tion is accepted (step 8), it is immediately refined with the same VND as above (step 9), making VND the workhorse also within ALNS. The destruction percentage adapts: it resets to a minimum after acceptance and increases after rejection (up to a cap), in order to broaden the explored neighborhood. Destroy operators include random removal; related removal, which selects customers similar to a random seed based on travel time and demand similarity; and worst removal, which targets customers with the highest marginal cost impact on the objective vector. Repair operators include greedy insertion, which inserts the customer with the smallest increase in objectives, and regret-k insertion, which computes the gap between the best and next-best k insertion costs and prioritizes largest regrets.

The final solution returned by ALNS, after systematic VND refinement of accepted states, constitutes the algorithm's output.

Algorithm 7. $ALNS\ (\hat{S}, \Omega^-, \Omega^+, LS)$

1: $S_{best} \leftarrow \hat{S}$
2: $S \leftarrow \hat{S}$
3: $initializeRouletteWheel()$
4: **while** not stopping criterion met **do**
5:     $destroy, repair \leftarrow wheelSelection(\Omega^+, \Omega^-)$
6:     $S' \leftarrow destroy(S)$
7:     $S'' \leftarrow repair(S')$
8:     **if** Criterion Acceptance **then**
9:         $S'' \leftarrow VND(S'', LS)$
10:         $S \leftarrow S''$
11:     **end if**
12:     **if** $isBetter(S'', S_{best})$ **then**
13:         $S_{best} \leftarrow S'''$
14:     **end if**
15:     $updateRouletteWheel()$
16: **end while**
17: **return** S_{best}

---

3   Experimental Results

This section presents the experimental results obtained for seven problem instances, each corresponding to a different city in Spain. The objective is to compare the performance of the proposed algorithms in realistic weekly planning scenarios. Due to confidentiality issues related to company product, the source code and data cannot be made publicly available.

In particular, for the solver-based algorithm, the maximum execution time was limited to 30 min for the model that considers the entire week globally. This limitation is based on the observation that most improvements in the solution are achieved during the first few minutes of the search. After approximately 10 to 15 min, additional improvements become progressively marginal. Therefore, a 30 min limit was set to ensure a proper balance between solution quality and computational cost.

It is important to remark that the problem being solved by Hexaly is a relaxation of the original problem, where problem constraints are transformed to objectives. More specifically, lateness, service consistency, vehicle overtime constraints are converted to objectives.

Additionally, the impact of parallelization was evaluated by varying the number of threads used. The thread count parameter represents how many parallel processes can run simultaneously during the algorithm's execution. A minimum of 4 threads was used, as recommended by the framework developers, and versions with 16 threads were also tested to observe scalability. It is worth mentioning that increasing the number of threads does not necessarily leads to better solutions. For instance, having more threads may lead the algorithm to stagnate in a different local optimum at the beginning of the search, which will prevent the algorithm to explore more promising regions of the search space

All algorithms were implemented in Python 3.12, and the experiments were executed on a machine with an 11th Gen Intel Core i5-1135G7 @ 2.40 GHz, 8 GB RAM and the value of the parameter α used in Algorithm 3 is set to $\alpha = 0.25$.

The columns in the tables contain the following information:

- **Alg.**: Name of the evaluated algorithm.
- **Delays**: Total accumulated delay time (in minutes).
- **Routes**: Total number of routes used.
- **T(min)**: Total driving time in minutes.
- **Op.**: Indicates whether the solution meets (≥ 90) or does not meet (< 90) the minimum service consistency requirement.
- **T(s)**: Execution time of the algorithm (in seconds). It is important to remark that, in a multi-threading scenario, this time is sized as the elapsed time between the start of the execution and its end, when a solution is provided.
- **Dev(%)**: Percentage deviation from the best known valid solution in the following order regarding the objectives: first, number of trucks is evaluated; when the number of trucks is the same for two different solutions, the accumulated delay value of both solutions is compared; when delay is the same, the total driving time is considered.
- **Best**: Indicates whether it is the best valid solution for the instance (marked with a 1).

To determine the best solution, a hierarchical criterion is followed based on the first three objectives of the problem:

1. First, the total delay is prioritized.
2. In case of a tie, the number of routes is compared.
3. If still tied, the total driving time is evaluated.

The objective of respecting vehicle capacity is not included, as all generated solutions satisfy this constraint. To facilitate reading, the best solution is highlighted in bold. Solutions that do not meet the minimum service consistency requirement (< 90) are not considered in the comparison process and are therefore not eligible as the "best solution". Each table corresponds to a city and shows the values obtained by each algorithm evaluated for that instance.

The instance in Table 1 corresponds to the city of Ávila. The maximum number of trucks available per day is 11, except on Saturdays, when the number is reduced to half (5 vehicles). The customer set includes a total of 418 locations, of which 90 require more than one visit per week.

The instance in Table 2 corresponds to the city of Rubí. The maximum number of trucks available per day is 13, except on Saturdays, when the number is reduced to half (6 vehicles). The customer set includes a total of 578 locations, of which 196 require more than one visit per week.

The instance in Table 3 corresponds to the city of Oviedo. The maximum number of trucks available per day is 8, except on Saturdays, when the number

Table 1. Results for the city of Ávila

Algorithm	Routes	Delays	T(min)	Op.	T(s)	Dev(%)	Best
Solver no solution	60	5532	30480	<90	1800	–	–
Solver solution 16 threads	62	62	25839	≥90	2724	3.33%	0
Solver solution 4 threads	60	47	25630	≥90	4545	193.75%	0
MH	**60**	**16**	**21568**	**≥90**	**2220**	**0%**	**1**

Table 2. Results for the city of Rubí

Algorithm	Routes	Delays	T(min)	Op.	T(s)	Dev(%)	Best
Solver no solution	68	0	30179	<90	1800	–	–
Solver solution 16 threads	69	9	25285	≥90	2565	4.55%	0
Solver solution 4 threads	**66**	**0**	**24839**	**≥90**	**3780**	**0%**	**1**
MH	67	0	25401	≥90	5010	1.52%	0

is reduced to half. The customer set includes a total of 374 locations, of which 99 require more than one visit per week.

The instance in Table 4 corresponds to the city of Barcelona. The maximum number of trucks available per day is 17, except on Saturdays, when the number is reduced to half (8 vehicles). The customer set includes a total of 1225 locations, of which 358 require more than one visit per week.

The instance in Table 5 corresponds to the city of Sevilla. The maximum number of trucks available per day is 14, except on Saturdays, when the number is reduced to half. The customer set includes a total of 527 locations, of which 262 require more than one visit per week.

The instance in Table 6 corresponds to a city near Barcelona. The maximum number of trucks available per day is 8, except on Saturdays, when the number is reduced to half. The customer set includes a total of 300 locations, of which 121 require more than one visit per week.

The instance in Table 7 corresponds to the city of Santander. The maximum number of trucks available per day is 12, except on Saturdays, when the number

Table 3. Results for the city of Oviedo

Algorithm	Routes	Delays	T(min)	Op.	T(s)	Dev(%)	Best
Solver no solution	44	713	19840	<90	1800	–	–
Solver solution 16 threads	44	0	16668	≥90	2580	6.12%	0
Solver solution 4 threads	44	0	16223	≥90	2760	3.29%	0
MH	**44**	**0**	**15707**	**≥90**	**1389**	**0%**	**1**

Table 4. Results for the city of Barcelona 1

Algorithm	Routes	Delays	T(min)	Op.	T(s)	Dev(%)	Best
Solver no solution	93	22497	50075	<90	1800	–	–
Solver solution 16 threads	93	649	39000	<90	4213	–	–
Solver solution 4 threads	93	464	38747	<90	4512	–	–
MH	**92**	**611**	**38541**	**≥90**	**24489**	**0%**	**1**

Table 5. Results for the city of Sevilla

Algorithm	Routes	Delays	T(min)	Op.	T(s)	Dev(%)	Best
Solver no solution	75	4008	34223	<90	1800	–	–
Solver solution 16 threads	75	0	27135	≥90	2784	0.36%	0
Solver solution 4 threads	**75**	**0**	**27038**	**≥90**	**2778**	**0%**	**1**
MH	75	0	28804	≥90	11900	6.53%	0

is reduced to half. The customer set includes a total of 355 locations, of which 162 require more than one visit per week.

In the comparison of experimental results, it was observed that in terms of solution quality, the use of metaheuristics achieved the best results in 3 out of the 7 evaluated instances (Ávila, Oviedo, and Barcelona 1), while the solver performed better in the remaining 4 cases. Regarding execution times, both approaches showed similar performance in small and medium-sized instances. However, for larger instances, the solver demonstrated significantly better computational efficiency.

A particularly noteworthy case is shown in Table 4, which corresponds to a highly complex instance due to its topology and the large number of customers. In this instance, the solver failed to generate an operational solution, whereas the metaheuristics succeeded in producing a feasible one, although with a considerably longer execution time (approximately 6 h and 48 min). This highlights the robustness of the proposed algorithm when tackling complex problems, while also emphasizing the importance of improving computational efficiency for large-scale instances.

Table 6. Results for Barcelona 2

Algorithm	Routes	Delays	T(min)	Op.	T(s)	Dev(%)	Best
Solver no solution	41	0	15541	≥90	1800	7.89%	0
Solver solution 16 threads	**38**	**0**	**13712**	**≥90**	**2272**	**0%**	**1**
Solver solution 4 threads	40	0	13767	≥90	2528	5.26%	0
MH	38	0	14075	≥90	948	2.65%	0

Table 7. Results for the city of Santander

Algorithm	Routes	Delays	T(min)	Op.	T(s)	Dev(%)	Best
Solver no solution	63	122	25812	<90	1800	–	–
Solver solution 16 threads	**61**	**0**	**20614**	**≥90**	**3126**	**0%**	**1**
Solver solution 4 threads	61	0	20826	≥90	3036	1.02%	0
MH	61	0	21568	≥90	1886	4.63%	0

4   Conclusions and Future Work

This work was developed in the context of a real-world weekly planning problem faced by a company that, until recently, relied on commercial software as its main tool for solving it. The solver had been adopted after a lengthy evaluation and development process that lasted over a year, during which its structure, configurations, and internal logic were adjusted to produce high quality solutions. Despite its effectiveness, the need arose to replace it with proprietary software, motivated by reasons of technological independence, full control over the optimization processes, and the ability to incorporate specific modifications more flexibly.

The main objective of this project was to carry out this transition to custom code without compromising the quality of the solutions. While the adaptation and exploitation of the solver took more than a year, the development of the new system was completed in less than two months. This not only demonstrates the efficiency of the chosen approach but also its potential for future extensions and improvements.

Given the limitation of number of pages, the preliminary experimentation for adjusting the final algorithm parameters is not reported in this work. Nevertheless, during the experimentation phase, various versions of the developed algorithms were compared with the solutions generated by the solver across different representative instances of the problem, each corresponding to a different city. The results are promising, as in several of these instances, the custom code matched or even outperformed the solver in the key problem criteria, especially in minimizing delays and routes, while always respecting operational constraints. Further testing is planned with larger instances, involving a greater number of customers and a more extensive network of cities. These tests will help confirm the scalability and robustness of the approach in more demanding scenarios.

One of the main areas for improvement identified is the need to reduce computation times so that the system can also compete with the solver in this aspect. Although the current design prioritizes solution quality, it is essential to also optimize computational efficiency to facilitate integration into a production environment. An immediate and straightforward improvement would be to parallelize the GRASP component of the algorithm, as its iterations are completely independent and can be executed in parallel, extracting only the best results at the end.

Additionally, it has been observed that customers with a visit frequency greater than one have a significant impact on solution quality. Therefore, a promising line of future work is to strengthen their treatment within the algorithm, both in the local search phase and in the destruction and repair strategies.

Another interesting line of future work is to approach the problem as a multi-objective optimization one. Transforming constraints that has direct implications on the quality of solutions will lead to an scenario in which the objective function to be optimized is compound by different conflicting-objectives.

In summary, this work has not only demonstrated that it is possible to replace solvers with a custom solution in a very short time frame, but it has also laid a solid foundation for a more powerful, adaptable, and efficient tool-aligned with the real needs of the company and with broad potential for future improvement and customization.

References

1. Flood, M.M.: The traveling-salesman problem. Oper. Res. **4**(1), 61–75 (1956)
2. Dantzig, G.B., Ramser, J.H.: The truck dispatching problem. Manage. Sci. **6**(1), 80–91 (1959)
3. Pullen, H., Webb, M.: A computer application to a transport scheduling problem. Comput. J. **10**(1), 10–13 (1967)
4. Beltrami, E.J., Bodin, L.D.: Networks and vehicle routing for municipal waste collection. Networks **4**(1), 65–94 (1974)
5. Groër, C., Golden, B., Wasil, E.: The consistent vehicle routing problem. Manufacturing & service operations management **11**(4), 630–643 (2009)
6. Rodríguez-Martín, I., Salazar-González, J.-J., Yaman, H.: The periodic vehicle routing problem with driver consistency. Eur. J. Oper. Res. **273**(2), 575–584 (2019)
7. Cordeau, J.-F.: The vrp with time windows. Groupe d'études et de recherche en analyse des décisions Montréal (2000)
8. Goeke, D., Roberti, R., Schneider, M.: Exact and heuristic solution of the consistent vehicle-routing problem. Transp. Sci. **53**(4), 1023–1042 (2019)
9. Rodríguez-Martín, I., Yaman, H.: Periodic vehicle routing problem with driver consistency and service time optimization. Transp. Res. Part B: Methodol. **166**, 468–484 (2022)
10. Braekers, K., Kovacs, A.A.: A multi-period dial-a-ride problem with driver consistency. Transp. Res. Part B: Methodol. **94**, 355–377 (2016)
11. Stavropoulou, F.: The consistent vehicle routing problem with heterogeneous fleet. Comput. Operat. Res. **140**, 105644 (2022)
12. Xu, Z., Cai, Y.: Variable neighborhood search for consistent vehicle routing problem. Expert Syst. Appl. **113**, 66–76 (2018)
13. Alegre, J., Laguna, M., Pacheco, J.: Optimizing the periodic pick-up of raw materials for a manufacturer of auto parts. Eur. J. Oper. Res. **179**(3), 736–746 (2007)
14. Kovacs, A.A., Golden, B.L., Hartl, R.F., Parragh, S.N.: The generalized consistent vehicle routing problem. Transp. Sci. **49**(4), 796–816 (2015)
15. Mladenović, N., Hansen, P.: Variable neighborhood search. Comput. Operat. Res. **24**(11), 1097–1100 (1997)

16. Feo, T.A., Resende, M.G.: A probabilistic heuristic for a computationally difficult set covering problem. Oper. Res. Lett. **8**(2), 67–71 (1989)
17. Pisinger, D., Ropke, S.: A general heuristic for vehicle routing problems. Comput. Operat. Res. **34**(8), 2403–2435 (2007)
18. Duarte, A., Mladenović, N., Sánchez-Oro, J., Todosijević, R.: Variable neighborhood descent. In: Handbook of Heuristics, pp. 1–27. Springer, Cham (2016)

Author Index

GPSR Compliance
The European Union's (EU) General Product Safety Regulation (GPSR) is a set
of rules that requires consumer products to be safe and our obligations to
ensure this.

If you have any concerns about our products, you can contact us on

ProductSafety@springernature.com

In case Publisher is established outside the EU, the EU authorized
representative is:

Springer Nature Customer Service Center GmbH
Europaplatz 3
69115 Heidelberg, Germany